CAMBRIDGE LIBRARY COLLECTION

Books of enduring scholarly value

Technology

The focus of this series is engineering, broadly construed. It covers technological innovation from a range of periods and cultures, but centres on the technological achievements of the industrial era in the West, particularly in the nineteenth century, as understood by their contemporaries. Infrastructure is one major focus, covering the building of railways and canals, bridges and tunnels, land drainage, the laying of submarine cables, and the construction of docks and lighthouses. Other key topics include developments in industrial and manufacturing fields such as mining technology, the production of iron and steel, the use of steam power, and chemical processes such as photography and textile dyes.

Practical Essay on the Strength of Cast Iron and Other Metals

Although cast iron was used in pagoda construction in ancient China, it was in Britain in the eighteenth century that new methods allowed for its production in quantities that enabled widespread use. An engineer who had educated himself tirelessly in technical subjects from carpentry to architecture, Thomas Tredgold (1788–1829) first published this work in 1822. It served as a standard textbook for British engineers in the early nineteenth century, and several translations extended its influence on the continent. Reissued here in the fourth edition of 1842, edited and annotated by the structural engineer Eaton Hodgkinson (1789–1861), who presents his own research in the second volume, this work addresses both practical and mathematical questions in assessing metallic strength. In Volume 1, wherever progress has been made since the original publication, Hodgkinson adds notes to Tredgold's original text, pointing out certain errors.

Cambridge University Press has long been a pioneer in the reissuing of out-of-print titles from its own backlist, producing digital reprints of books that are still sought after by scholars and students but could not be reprinted economically using traditional technology. The Cambridge Library Collection extends this activity to a wider range of books which are still of importance to researchers and professionals, either for the source material they contain, or as landmarks in the history of their academic discipline.

Drawing from the world-renowned collections in the Cambridge University Library and other partner libraries, and guided by the advice of experts in each subject area, Cambridge University Press is using state-of-the-art scanning machines in its own Printing House to capture the content of each book selected for inclusion. The files are processed to give a consistently clear, crisp image, and the books finished to the high quality standard for which the Press is recognised around the world. The latest print-on-demand technology ensures that the books will remain available indefinitely, and that orders for single or multiple copies can quickly be supplied.

The Cambridge Library Collection brings back to life books of enduring scholarly value (including out-of-copyright works originally issued by other publishers) across a wide range of disciplines in the humanities and social sciences and in science and technology.

Practical Essay on the Strength of Cast Iron and Other Metals

Containing Practical Rules, Tables, and Examples, Founded on a Series of Experiments, with an Extensive Table of the Properties of Materials

VOLUME 1:
PRACTICAL ESSAY ON THE STRENGTH OF CAST IRON AND OTHER METALS

THOMAS TREDGOLD
EDITED BY EATON HODGKINSON

CAMBRIDGE
UNIVERSITY PRESS

University Printing House, Cambridge, CB2 8BS, United Kingdom

Cambridge University Press is part of the University of Cambridge.
It furthers the University's mission by disseminating knowledge in the pursuit of education, learning and research at the highest international levels of excellence.

www.cambridge.org
Information on this title: www.cambridge.org/9781108070348

This edition first published 1842
This digitally printed version 2014

ISBN 978-1-108-07034-8 Paperback

PRACTICAL ESSAY

ON THE

STRENGTH OF CAST IRON

AND OTHER METALS:

CONTAINING

PRACTICAL RULES, TABLES, AND EXAMPLES, FOUNDED ON A SERIES OF EXPERIMENTS;

WITH AN EXTENSIVE

TABLE OF THE PROPERTIES OF MATERIALS.

BY THOMAS TREDGOLD,

MEMBER OF THE INSTITUTION OF CIVIL ENGINEERS,

AUTHOR OF 'THE HISTORY OF THE STEAM ENGINE,' 'ELEMENTARY PRINCIPLES OF CARPENTRY,' &c., &c.

THE FOURTH EDITION, WITH NOTES BY

EATON HODGKINSON, F.R.S.

TO WHICH ARE ADDED

EXPERIMENTAL RESEARCHES

ON THE

STRENGTH AND OTHER PROPERTIES OF CAST IRON;

WITH

THE DEVELOPEMENT OF NEW PRINCIPLES; CALCULATIONS DEDUCED FROM THEM; AND INQUIRIES APPLICABLE TO RIGID AND TENACIOUS BODIES GENERALLY.

BY THE EDITOR.

LONDON:

JOHN WEALE, 59, HIGH HOLBORN.

MDCCCXLII.

TO

JAMES WALKER, ESQ.,

CIVIL ENGINEER,

FELLOW OF THE ROYAL SOCIETY,
&c., &c., &c.,

PRESIDENT OF THE INSTITUTION OF CIVIL ENGINEERS,

This Essay

ON THE

STRENGTH OF CAST IRON AND OTHER METALS,

RE-EDITED, WITH NOTES, BY

EATON HODGKINSON, ESQ., F.R.S.,

IS INSCRIBED BY

THE PUBLISHER.

PART I.

PRACTICAL ESSAY

ON

THE STRENGTH OF CAST IRON

AND OTHER METALS.

BY THOMAS TREDGOLD.

ADVERTISEMENT TO THE FOURTH EDITION.

IN giving a Fourth Edition of TREDGOLD'S 'ESSAY ON THE STRENGTH OF CAST IRON,' I have made no alteration in the text, but left it as it was in the last edition. The object of the very ingenious Author was to consider the resistances of bodies subjected to small forces, when compared with those necessary to break them; since with the action of small forces, the displacement of the fibres, or particles of bodies, is equal, from equal forces, whether they produce extension or compression. But conclusions drawn from such small strains, when applied to measure the ultimate strength of cast iron, are often much at variance with the results of experiment.

An instance or two may be mentioned: a cast iron beam, to sustain most efficiently a moderate strain, should have equal ribs at top and bottom; but to offer the greatest resistance to fracture, these ribs should be unequal in the proportion of seven to one, nearly; and by this form a great addition of strength is obtained.

Under small strains beams of any particular form offer the same resistance, whether they are turned their proper way up or the reverse; but I have shown that a cast iron beam may be constructed to resist fracture with four times as much force one way up as the opposite.

The line bounding the extended and compressed fibres of a bent beam, called the neutral line, is in the middle of a square beam subjected to a small strain; but in a cast iron beam of this form, the neutral line has, at the time of fracture, removed near to the compressed side, and the strength is considerably increased by the change.

These results, with respect to fracture, arise from the circumstance that cast iron resists fracture in crushing with many times the force that it does in tearing asunder; the mean being about seven times, nearly.

The preceding facts show an essential difference in the laws which regulate moderate and ultimate strains; and the latter will be considered in the Second Part of this Work.

The experiments from which Mr. Tredgold had to draw his conclusions, respecting the transverse strength of cast iron, had not been observed with sufficient accuracy to enable him to determine when the elasticity first became injured; and accordingly he concludes that beams retain their elasticity unimpaired till nearly one-third of the breaking weight is laid on; but I have shown that beams, whose section is **T**, and which bear so much more one way up than the other, will retain a perceptible deflexion, and not recover their original form, after as little as $\frac{1}{52}$d, or even $\frac{1}{80}$th of the breaking weight is laid on. In other words, there is no elastic limit, a set taking place with the smallest flexure. From this cause, and others, Mr. Tredgold has drawn some erroneous conclusions, of which a few are pointed out in the Notes to the body of his Work.

In a material so much used as cast iron, it is of great consequence to the founder to know whence he can obtain the irons best suited for different purposes. The experiments of Mr. Fairbairn on the transverse strength of cast iron bars, of which I have given an abstract, contain, with a few experiments made by myself, examinations, similarly conducted, of most of the irons used in this country: they will be consulted with interest for the purpose above mentioned; and his experiments on bars loaded for an indefinite time show that cast iron may be trusted far beyond what has generally been conceived.

The strength of pillars is a subject on which there has confessedly been a great want of experimental information: I therefore recently undertook an extensive series of experiments upon pillars, comparing the results with the theoretical conclusions of Euler and Lagrange upon the matter. With the reception by the Royal Society of the Paper containing these experiments, and still more with the honourable mark of distinction awarded to it, I have the fullest reason to be gratified; and I am indebted to the Council for the privilege of giving an abstract of it in this Work.

EDITOR.

ADVERTISEMENT TO THE THIRD EDITION.

PUBLIC approbation of 'THE PRACTICAL ESSAY ON THE STRENGTH OF CAST IRON, &c.' having made a new Edition necessary, it may be proper to state, that it is printed from a copy corrected by Mr. TREDGOLD's own hand, and to which nothing has been added; its progress through the press has been kindly superintended by the Author's friend, Professor Barlow, of Woolwich; its correctness cannot therefore be reasonably questioned.

The utility and importance of this Practical Essay have been acknowledged by the most unqualified approbation of the scientific of all countries, and it has accordingly been printed in the French, Italian, and Dutch languages. To the clear and practical demonstrations set forth in this Work, of the superiority of iron for supports, as well perpendicular as horizontal, may be attributed the present almost universal adoption of this material in buildings, as a substitute for wood; and, in reference to this particular subject, the Author was frequently consulted by the most eminent Architects and Engineers. His many other valuable works, of which a list is subjoined, more particularly the 'History of the Steam Engine,' have also experienced a similar liberal patronage both at home and abroad; and it is not, we think, presuming too much to say, that these Works have had an important influence in promoting the present advanced state of mechanical and scientific knowledge,—and to this honour the Author ardently aspired, as he considered it of the highest value.

Of Mr. Tredgold it may be stated, that from his earliest

years his mind was ever occupied by the most intense desire for information. Being altogether *self-taught*, it will not be difficult to form some idea of the great labour and fatigue which he must have necessarily undergone, in the acquirement of that correct and extensive knowledge of the various sciences of which he has so ably treated; and in his high attainments in mathematical demonstration, which he has so ingeniously and successfully applied in the many useful investigations exhibited in his Works.

To such ardent and unremitted application, the naturally feeble constitution of Mr. TREDGOLD eventually gave way, and after a protracted series of suffering terminated in his death on the 28th of January, 1829, in the 41st year of his age,—to the great loss of the public at large, and which an amiable wife and young family have seriously to deplore;[1] whose slender circumstances call

---

[1] The widow lived but a short time after the death of her husband, leaving three daughters and a son on the scanty subsistence afforded by a subscription set on foot by some benevolent Members of the Profession of Civil Engineering and Architecture: subsequently the two elder daughters died. The son is now articled to Mr. Bryan Donkin, of Rotherhithe, and the only daughter is living under the protection of Mrs. Urquhart, her aunt, who is herself in straitened circumstances from the recent loss of her husband, on whom she wholly depended for support. At no time nor in any place can it be improper, in the cause of humanity, to mention the fact that Mrs. Urquhart, as a measure of relief from the heavy burden upon her hand, is now endeavouring to procure a presentation for one of her children to Christ's Hospital. Besides the strong claim before referred to, she has the following certificate from Dr. Reid, who testifies to her respectability and merit:

"I hereby certify that the late John Urquhart, who attended to the Warming and Ventilating of the present Houses of Parliament under my direction, conducted himself with the greatest propriety, steadiness, and attention to all the duties intrusted to

for the kind aid of all benevolent persons, in behalf of these representatives of a truly worthy and scientific man, who devoted his whole life to the promotion and general diffusion of the most useful knowledge, with scarcely an ordinary attention to his own personal emolument.

As a philosopher and author, Mr. TREDGOLD contributed to many of the scientific publications of the day, to which his name was not always added. The following List of his Works will serve to evince his great industry, and will show the extensive range of studies with which his mind was successively occupied, and the important results he ever had in view:

---

him, and that the zeal and intelligence with which he directed the operations he had to conduct gave the highest satisfaction. In the prime of his life, and when he was beginning to emerge from the difficulties with which he had long struggled, he was suddenly cut off, and has left a wife and family to deplore his loss. From the high character which he always had, and from the difficulties and embarrassments in which his widow (a near relative of the late Mr. Tredgold, to whose family she still devotes the care and anxiety of a mother,) is necessarily placed, and the exemplary manner in which she continues to bear up against the adversity with which she has been nearly overwhelmed, I have thought it incumbent on me to bear testimony to the circumstances I have now stated, and shall merely add that her case demands the sympathy and consideration of those who have the means of assisting her, and who will find on inquiry that her conduct has been such as to call for the most liberal assistance that her wants and necessities require.

"D. B. REID."

It is therefore hoped that some benevolent individual will endeavour to procure from a trustee or subscriber the required presentation; and in the true spirit of a grateful remembrance of the talent of the late Mr. Tredgold, who did so much for his succeeding age, this case is earnestly placed before the Architectural and Engineering Profession by the Publisher.—1842.

1. Elementary Principles of Carpentry, 4to. 22 Plates. (Re-edited by Professor BARLOW in 1838, and containing 28 additional Plates and an Appendix.)
2. A Treatise of Joinery, in the Encyclopædia Britannica.
3. Essay on the Strength of Cast Iron and other Metals, 8vo. 4 Plates.
4. Additions to Buchanan's Essays on Mill-work, 2 vols. 8vo. 20 Plates. (Re-edited, with much additional matter, principally treating of Tools, Machines, &c., by GEORGE RENNIE, C. E., F.R.S. 74 Plates, in 2 vols.)
5. A Treatise on Stone Masonry, for Supplement to Encyclopædia Britannica.
6. Principles of Warming and Ventilating Public Buildings, Hot-houses, &c., 8vo. 9 Plates. (Re-edited by Mr. BRAMAH in 1838.)
7. A Treatise on Rail-roads and Carriages, &c., 8vo. 4 Plates.
8. A Letter to Mr. Huskisson on Steam Navigation, 8vo.
9. Additions and Notes to Tracts on Hydraulics, by SMEATON, VENTURI, Dr. YOUNG, &c. 8vo. 7 Plates.
10. Practical Rules, with Diagrams, for BARLOW's Essay on the Strength of Timber, 8vo.
11. The Steam Engine; comprising an account of its Invention, Progressive Improvement, &c., 4to. 20 Plates. (This Work has been much extended to all the practical improvements of the last ten years, and particularly as regards STEAM NAVIGATION and LOCOMOTIVE ENGINES, in 2 large volumes, with 125 Plates. 2300 copies of this edition have been sold in three years.—PUBLISHER.)

On recording Mr. TREDGOLD's death, the Editor of the Literary Gazette well observes, "The numerous and excellent publications of Mr. TREDGOLD will ever hold the first place among the elementary compendiums of Civil Engineering, and must ever insure him the lively gratitude of the cultivators of general knowledge."

J. T.

August, 1831.

PREFACE TO THE SECOND EDITION.

In the following pages I have attempted to supply a 'Practical Treatise on the Strength of Cast Iron:' the use and advantage of such a Work will be best appreciated by those who consider the serious consequences of a failure in the application of this material. It is used for the principal supports of Churches, Theatres, Dwelling-houses, Manufactories, and Warehouses; for Bridges, Roofs, and Floors; and for the moving parts of the most powerful Engines. If a failure take place, from want of strength, it will most probably happen at that moment when its consequences will be most serious: hence, I think I may venture to say, without giving any undue importance to the object of this Work, that, if there be one subject which requires the aid and assistance of science more than another, it is the application of supports of cast iron.

The very considerable improvements that have been made in the manufacture of iron, have, undoubtedly, chiefly arisen out of the peculiar advantages derived from its use, in the mining and manufacturing districts of Britain; and the immense quantities of it employed in these districts is one of the most satisfactory proofs of its utility and value.

These improvements in the manufacture of iron have

also enabled the manufacturers to reduce its price; so far indeed that it now can be employed, instead of foreign timber, for many important purposes in buildings and machines, at a very small additional expense, with a considerable addition of soundness and durability. It is not, however, fitted for every purpose; for example, if it be desirable that a house should exclude the cold of winter, and the heat of summer, it certainly would not be adviseable to form the roof, or any other considerable part of it, wholly of iron; as you could not easily find another substance for the purpose, that suffers heat to pass through it so rapidly as iron does. But it is more imprudent to build heavy brick or stone walls upon timber supports, a material which is so subject to decay, and so easily destroyed by fire; and yet nearly half the houses in London are partly sustained by wooden posts. If you use timber to prevent settlements where a foundation is soft or irregular, the timber decays, and worse settlements take place than those it was intended to avoid:[1] in all such cases iron might be used with success.

I think it will appear, on an accurate survey of the present state of the mechanical arts, that the physical and mechanical properties of matter are not sufficiently studied. If such knowledge were cultivated, if it formed a part of a young mechanic's education,—that is, if he were prepared by a regular course of experimental study respecting the nature and properties of materials,—would not his progress in any particular art be greatly facilitated? Experience, "slow preceptress," furnishes a practical mechanic with some share of this knowledge, but such experience is always limited to a particular

---

[1] See Napier's Supplement to Encyclo. Brit. art. 'Stone Masonry,' § 60.

range of objects, and it engenders prejudice in favour of particular things, and particular modes of operation. Lord Bacon's idea of a Mechanical History,[2] which Diderot attempted to realize,[3] is not so well calculated to fulfil his own view as a well-directed course of experiments on the nature, forms, and properties of materials, illustrated by a reference to the manner of applying them in the arts. In Chemistry much has been already done; but an Experimental School of Mechanical Science remains to be formed.

Having briefly alluded to this deficiency in the experimental investigation of the mechanical properties of bodies, I must proceed to inform the reader of the nature of the Work now offered to his notice, as improved by the addition of another year's collection of experience, and much experimental research.

This Work is divided into eleven Sections:—The First Section consists of introductory remarks on the use and the qualities of cast iron; and of cautions to be observed in employing it. This section includes three extensive Tables, which will often save the practical man a considerable share of trouble in calculation.

The Second Section explains the arrangement and use of the Tables which precede it; and in this edition the number of popular examples is much increased.

It is a common and a well understood fact, that an uniform beam is not equally strained in every part, and therefore may be reduced in size, so as to lessen both the strain and the expense of material.

The Third Section points out the value of cast iron in

---

[2] Of the Advancement of Learning. Book II. Bacon's Works, vol. i. p. 79.

[3] French Encyclopédie.

this particular, and the forms to be adopted for different cases.

The Fourth Section contains a popular explanation of the strongest forms for the sections of beams; the construction of open beams; and the best form for shafts. A due consideration of these two sections will enable the young mechanic to guard against some common errors in attempting to apply these things to practice. They are much augmented, and a new principle of constructing bridges is explained in the fourth section.

The Fifth Section is wholly devoted to experiments on cast iron: it will be found to contain, in addition to my own experiments, almost all of the experiments that have been described by preceding writers. Those I have tried for the purpose of establishing rules, to apply in practice, have been made with a different view of the subject from that entertained by preceding experimentalists; one better adapted for practical application, one which shows that, within the proper limits, our theory of the strength of materials is to be depended upon; but that beyond these limits materials should never be strained in constructions of any kind whatever.[4] Nevertheless it would be extremely desirable that some accurate experiments on the extension of bodies should be made, when the strain exceeds the elastic force; as by that means something important regarding the ductility of matter might be discovered; and perhaps they might throw some light on the nature and arrangement of the ultimate particles of bodies.

---

[4] To Dr. T. Young we are chiefly indebted for showing the necessity of attending to the strain which produces permanent alteration: Nat. Phil. vol. i. p. 141. To that valuable Work I am most indebted for assistance in this Essay.

To this section a great many new experiments have been added, to show the relative strength of iron of different qualities; and also seven new experiments on torsion, made by Messrs. Bramah. The section concludes with the result of my observations on the relation between the appearance of the fracture and the strength of cast iron, as determined by experiment.

The Sixth Section contains experiments on malleable iron and other metals, and is entirely new. The effect of hammering, and the decrease of force by heat, are experimentally examined; and the cause of English iron being inferior to Swedish, for particular purposes, is pointed out.

In the Seventh Section I have shown how to obtain some of the most useful practical rules from the first principles that are furnished by experience. I have conducted the investigation of these rules in a manner somewhat different from other writers, and I have avoided the use of fluxions.[5] Several new cases are investigated,

[5] I have avoided fluxions in consequence of the very obscure manner in which its principles have been explained by the writers I have consulted on the subject. I cannot reconcile the idea of one of the terms of a proportion vanishing for the purpose of obtaining a correct result; it is not, it cannot be good reasoning; though, from other principles, I am aware that the conclusions obtained are accurate. If the doctrine of fluxions be freed from the obscure terms, limiting ratios, evanescent increments and decrements, &c., it is in reality not very difficult. If you represent the increase of a variable quantity by a progression, (as is done in art. 295, Section XI.,) any term of that progression (except the last) corresponds with what is called a fluxion; and the sum of the progression is the same as a fluent. A fluxion is, therefore, the velocity of increase of an increasing variable quantity; or of decrease of a decreasing one; on the supposition that we take the velocity at any point and consider it uniform. But

and some addition is made to the theory of resistance: the reader will find examples of this in treating of the strength of beams, art. 108 to 119; the deflexion of beams, art. 124 to 130; the strain upon beams, art. 133 to 141; and in the ninth and tenth sections.

The Eighth Section treats of the stiffness to resist lateral strains, with its application to some interesting practical cases.

The Ninth Section is on the strength and stiffness

---

when you use this uniform velocity to represent an accelerated one, and say that the ratio of these velocities approaches to a ratio of equality as its limit, while the space or time of description is diminishing to evanescence, I must be allowed to withhold my assent to your doctrine. For it is clear that the exact ratio of equality can obtain only when the space described is nothing; and consequently, it cannot, with logical accuracy, be employed to compare the spaces generated when they become of finite magnitude. Robins and Maclaurin have shown that their reasoning is agreeable to the practice of the ancient geometers; but to give dignity to their speculations these geometers employed a subtile and metaphysical method, in preference to a candid avowal of a tentative comparison: the same practice there is not now any reason to continue. The science of space, or geometry, is completely distinct from that of number or analysis. They have both been injured by an intermixture, which was begun by the ancients: how they came to conceive that the doctrines of number and space rested on principles common to both, it is not necessary to inquire; but as soon as you commence the fifth book of Euclid you bid adieu to pure geometry; and, in the rest, the aid of a metaphysical method of forcing the assent, rather than of convincing the judgment, is frequently introduced, and reigns through most of the works of the older geometers. The neglect of pure geometry has left ample scope for a display of talent. I am thankful to the kind friends who have taken some pains to correct my notions on this subject; they will see that I have profited in some degree by their remarks.

to resist torsion or twisting, with its application to machinery.

The Tenth Section treats of the strength of columns, pillars, and ties, with some new examples. It may be useful to remark, that the most refined methods of analysis have been applied to the same subjects by Euler, Lagrange, and other continental mathematicians, without arriving at results more accurate, more simple, or more convenient in practice.

In the Eleventh Section I have considered the resistance of beams to impulsive force. In this section will be found many important rules, with examples of their application to the moving parts of engines, bridges, &c., wherein the advantage gained by employing beams of the figures of equal resistance is shown.

The Eleventh Section is followed by an extensive 'TABLE OF THE PROPERTIES OF MATERIALS, AND OTHER DATA, OFTEN USED IN CALCULATIONS,' arranged alphabetically, and in this edition much enlarged. By means of this Table the various rules for the strength of cast iron, contained in this Work, may be applied to several other kinds of materials.

A note, which I have added at the end of the Table, on the chemical action of some bodies on cast iron, will be read with interest by those who employ cast iron where it is exposed to the action of sea-water.

The Plates are accompanied by descriptive letter-press, with references to the articles which the figures are intended to explain.

The Work concludes with an Index,[6] containing copious references to the Practical Rules; and, in general, it will be found that the examples are selected with a view to

---

6 A *General Index* will be given at the end of the Work.—ED.

explain the practical application of the rules, and to make the reader aware of the limits and precautions to be attended to. In fact, the want of such information has often brought theory into discredit with some men, whereas the fault ought to have fallen on the person that misapplied it.

I hope there will be few things of any importance found in this Work for which a sufficient reason is not given: sometimes I have been compelled to omit several steps in the investigations, in order to make it as little mathematical as possible; and such omissions the reader must excuse till a larger share of mathematical learning becomes the common lot of every practical mechanic; and I hope that period is not far distant.

The communication of any experiment or observation that is calculated to confirm or correct any thing I have done I shall esteem a favour.

The manner in which my Works have been received has been highly gratifying to my feelings, and has afforded me an early opportunity of rendering my grateful thanks. They are especially due to Messrs. Bramah and Mr. Bevan,—the former, for specimens and an account of experiments,—the latter, for corrections of some press errors.

A Second Part is in progress, on the 'Strength of Pipes, Mains, Tanks, Boilers, &c.; of Chains to resist Impulsion and Pressure; of Suspension and other Iron Bridges; and of Framed Work for Roofs, Bridges, Mills, and Machinery.'

16, *Grove Place, Lisson Grove,*
Nov. 1823.

CONTENTS OF PART I.

SECTION I.

SECTION II.

SECTION III.

SECTION IV.

SECTION V.

SECTION VI.

SECTION VII.

SECTION VIII.

SECTION IX.

SECTION X.

SECTION XI.

All the quantities, proportions, &c., are stated in English weights and measures, the pound being the avoirdupois pound, except the contrary be stated.

SECTION I.

INTRODUCTION.

Art. 1. In consequence of the security which cast iron gives, when it is properly employed, for supporting considerable weights, pressures, or moving forces, it has lately been very much used; and is likely to wholly supersede the use of timber for many important purposes. Indeed, so considerable are the improvements which have arisen out of its use, that the period of its general introduction has been very justly considered as forming a new era in the history of machines.[1] "All other improvements," it has been remarked, "have been limited; confined to particular machines; but this, having increased the strength and durability of every machine, has improved the whole."[2]

Cast iron is a valuable material, because it gives

[1] Essays on Mill Work, &c., by Robertson Buchanan, Essay II. p. 254, 2d edit.; or 3rd edit. by G. Rennie, Esq., p. 177.

[2] Mr. Dunlop's Account of some Experiments on Cast Iron. Dr. Thomson's Annals of Philosophy, vol. xiii. p. 200.

safety against fire; it is not liable to sudden decay, nor soon destroyed by wear and tear, and it can be easily moulded into the form of greatest strength, or that which is best adapted for our intended purpose.

The fatal consequences that might result from the use of timber for supporting heavy buildings, either in case of fire or of decay, have often been foreseen; but in a few instances it has happened that where iron has been used for greater security against fire, the structure has failed from want of strength. Such failures have not occurred from any defect in the material itself; for it too often happens that such works are conducted by persons of little experience, and less scientific knowledge. Men of little experience too frequently imagine that a large piece of iron is almost of infinite strength; and they often have a like indistinct notion of pressure. They design to please the eye, without regard to fitness, strength, or durability; instead of ornamenting a support, they make the support itself the ornament, and sacrifice every thing to lightness of effect. The dimensions of the most important parts of structures are too often fixed by guess or chance; and the person who calculates the value of materials to the fraction of a penny, seldom if ever attempts to estimate their power, or the stress to which they will be exposed.

The manner in which the resistance of materials has been treated by most of our common mechanical writers, has also, in some degree, misled such practical men as were desirous of proceeding upon surer ground; and has given occasion for the sarcastic remark, "that the stability of a building is inversely proportional to the science of the builder." [3]

When it is considered that it is absolutely necessary that the parts of a building or a machine should preserve a certain form or position, as well as that they should bear a certain stress, it will become obvious that something more than the mere resistance to fracture should be calculated. In cases where the parts are short and bulky, it may do very well to employ the rules for resistance to fracture, and make the parts strong enough to sustain four times the load, but such cases rarely occur; and where long pieces are loaded to one-fourth of their strength, we may expect much flexure, vibration, and instability.

If a material of any kind be loaded with more than a certain quantity, it loses the power of recovering its natural form, when the load is removed; the arrangement of its particles undergoes a permanent alteration; and if it supports the same load during a considerable time, the deflexion will increase, and the more in proportion as the load is above the elastic force of the material.[4]

---

[3] Ency. Method. Dict. Architecture, art. Equilibre.

[4] This important fact appears to have been first noticed by Coulomb, while making his experiments on torsion. (Some account has been given of Coulomb's experiments by Dr. Young,

On this part of the resistance of materials I have made many experiments, both with metals of various kinds, and with timber: I find that while the elastic force or power of restoration remains perfect, the extension is always directly proportional to the extending force, and that the deflexion does not increase after the load has been on for a second or two; but when the strain exceeds the elastic force, the extension or deflexion becomes irregular, and increases with time. I was led into this important inquiry by considering the proportions for cannon,

---

Nat. Phil. vol. ii. p. 383, and also by Dr. Brewster in his additions to Ferguson's Lectures, vol. ii. p. 234, third edit.) But, in a great number of substances, we seem to have an instinctive knowledge of this property of matter: a bent wire retains its curvature; and it may be broken by repeated flexure, with much less force than would break it at once: indeed, when we attempt to break any flexible body, it is usually by bending and unbending it several times, and its strength is only beyond the effort applied to break it when we have not power to give it a permanent set at each bending. A permanent alteration is a partial fracture, and hence it is the proper limit of strength. Dr. Young, with his usual profound discrimination, pointed out the importance of this limit in applying the discoveries in science to the useful arts.

While I was preparing this edition for the press, I received a copy of the "Essai Théorique et Expérimental sur la Résistance du Fer Forgé," of M. Duleau, which is founded on similar views of the strength of wrought iron. M. Duleau has ascertained, with an apparatus much more imperfect than mine, the fact that iron cannot be considered a perfectly elastic body when the strain exceeds a certain force. I shall, in the course of this edition, compare the results of his experiments with those I have made, wherever the conditions are similar.

and the common method of proving them. It appears from my experiments, that firing a certain number of times with the same quantity of powder would burst a cannon when the strain is above the elastic force of the material, though the effect of the first charge might not be sensible. The same remarks apply to the methods of proving the strength of steam engine boilers and pipes, by hydraulic pressure: if the strain in proving exceeds that which produces permanent alteration, an irreparable injury is done by the trial.

In the moving parts of machines the strain should obviously be under the elastic force of the material, and in the second Table will be found the flexure and load a piece of a given size will bear without destroying the elastic force.

I think every one, who carefully examines the subject, will feel satisfied that the measure of the resistance of a material to flexure is the only proper measure of its resistance, when it is to be applied where perfect form or unalterable position is desirable; and the measure of its resistance to permanent alteration, when it is used where flexure is not injurious nor objectionable.

In order to supply practical men with a convenient and ready means of assigning the dimensions of cast iron beams, columns, &c., to support known pressures, or moving forces, I have drawn up this volume. I am persuaded that its usefulness will find it a place among the common works of reference, which are more or less necessary to every architect, engineer, and builder. To bring it within as small a compass as possible, I have arranged the Tables so as to include as many distinct applications as the nature of the subjects seemed capable of admitting.

SOME PARTICULARS TO BE OBSERVED IN USING THE TABLES.

2. The weight of the beam itself is always to be estimated, and added to the load to be supported; or (because this method renders it necessary to estimate the weight before the bulk be determined) find the dimensions of the piece that would support the load by one of the Tables, and increase the breadth in the same proportion as the weight of the piece increases the load. If the weight of the piece, for example, be an eighth part of the load, then to the breadth, found by the Table, add an eighth part of that breadth; and so of any other proportion. It is not an absolutely correct method, but it is simple and correct enough for use.

3. The Tables and Rules are calculated for soft gray cast iron. Metal of this kind yields easily to the file when the external crust is removed, and is slightly malleable in a cold state. Dr. C. Hutton has justly given the preference to such iron, because it is "less liable to fracture by a blow, or shock, than the hard metal." [5]

---

[5] Tracts, vol. i. p. 141.

White cast iron is less subject to be destroyed by rusting than the gray kind; and it is also less soluble in acids; therefore it may be usefully employed where hardness is necessary, and where its brittleness is not a defect; but it should not be chosen for purposes where strength is necessary. When it is cast smooth, it makes excellent bearings for gudgeons or pivots to run upon, and is very durable, having very little friction.

White cast iron, in a recent fracture, has a white and radiated appearance, indicating a crystalline structure. It is very brittle and hard.

Gray cast iron has a granulated fracture, of a gray colour, with some metallic lustre; it is much softer and tougher than the white cast iron.

But between these kinds there are varieties of cast iron, having various shades of these qualities; those should be esteemed the best which approach nearest to the gray cast iron.

Gray cast iron is used for artillery, and is sometimes called gun-metal.

The best and most certain test of the quality of a piece of cast iron, is to try any of its edges with a hammer; if the blow of a hammer make a slight impression, denoting some degree of malleability, the iron is of a good quality, provided it be uniform: if fragments fly off, and no sensible indentation be made, the iron will be hard and brittle.[6]

[6] For more information upon this subject, see Mr. Fairbairn's

The utmost care should be employed to render the iron in each casting of an uniform quality, because in iron of different qualities the shrinkage is different, which causes an unequal tension among the parts of the metal, impairs its strength, and renders it liable to sudden and unexpected failures. When the texture is not uniform, the surface of the casting is usually uneven where it ought to have been even. This unevenness, or the irregular swells and hollows on the surface of a casting, is caused by the unequal shrinkage of the iron of different qualities. A founder of much observation and experience in his business, pointed out to me this test of an imperfect casting.

Now, when iron of a particular quality is obtained by mixture of different kinds, it will be difficult to blend them so thoroughly as to render the product perfectly uniform; hence we easily perceive one reason of iron being improved by annealing, for in passing slowly to the solid state, the parts are more at liberty to adjust themselves, so as to equalize, if not neutralize, the tension produced by shrinking. But, it is clear that an annealing heat applied after the metal has once acquired its solid state, must be sufficiently intense to reduce the cohesive power in a very considerable degree, otherwise it will not be

---

Experiments upon the Transverse Strength, &c. of Bars of Cast Iron, from various parts of the United Kingdom. (Manchester Memoirs, vol. vi. new series).—EDITOR.

sensibly beneficial.[7] These remarks apply to glass, and to various metals as well as to cast iron.

It has been remarked that "iron varies in strength, and not only from different furnaces, but also from the same furnace and the same melting; but this seems to be owing to some imperfection in the casting, and in general iron is much more uniform than wood."[8] I am glad to find my own experience supported by the opinion of a writer so well known to practical men as Mr. Banks. But the very great strain which large masses of well mixed cast iron will bear, when applied to resist the greatest stresses in mill and engine work, is now extremely well known in this country. Its value was foreseen by our celebrated Smeaton at an early period of his practice. Upwards of forty years ago he combated the prejudices against it in the following language: "If the length of time of the use of these (cast iron) utensils is not thought sufficient, I must add, that in the year 1755, that is, twenty-seven years ago, for the first time, I applied them as totally new subjects, and the cry then was, that if the strongest timbers are not able for any great length of time to resist the action of the powers, what must happen from the brittleness of cast iron? It is sufficient to

---

[7] Dr. Brewster has shown that the mechanical condition of unannealed glass is not capable of being altered by the heat of boiling water. Edin. Phil. Journal, vol. ii. p. 399.

[8] Banks on the Power of Machines, p. 73. See also p. 94 of the same work.

say, that not only those very pieces of cast iron are still at work, but that the good effect has in the north of England, where first applied, drawn them into common use, and I never heard of one failing."[9] These remarks were written in 1782, and the good opinion of Smeaton has been fully justified by the experience of succeeding engineers; the grand and varied works of Wilson, Rennie, Boulton and Watt, Telford, &c., &c., abundantly confirm it.[10]

Yet I must not omit to remark, that cast iron

---

[9] Reports, vol. i. pp. 410, 411.

[10] One of the boldest attempts with a new material was the application of cast iron to bridges: the idea appears to have originated, in the year 1773, with the late Thomas Farnolls Pritchard, then of Eyton Turrett, Shropshire, architect, who, in communication with the late Mr. John Wilkinson, of Brosely and Castlehead, ironmaster, suggested the practicability of constructing wide iron arches, capable of admitting the passage of the water in a river, such as the Severn, which is much subject to floods. This suggestion Mr. Wilkinson considered with great attention, and at length carried into execution between Madely and Brosely, by erecting the celebrated iron bridge at Colebrook Dale, which was the first construction of that kind in England, and probably in the world. This bridge was executed by a Mr. Daniel Onions, with some variations from Mr. Pritchard's plan, under the auspices and at the expense of Mr. Darby and Mr. Reynolds, of the iron works of Colebrook Dale. Mr. Pritchard died in October, 1777. He made several ingenious designs, to show how stone or brick arches might be constructed with cast iron centres, so that the centre should always form a permanent part of the arch. These designs are now in the possession of Mr. John White, of Devonshire Place, one of his grandsons, to whom I am indebted for the preceding particulars of this note.

when it fails gives no warning of its approaching fracture, which is its chief defect when employed to sustain weights or moving forces; therefore care should be taken to give it sufficient strength. And it will be obvious, from the preceding remarks, how much its strength depends upon the skill and experience of the founder.

4. The parts of each casting should be kept as nearly of the same bulk as possible, in order that they may all cool at the same rate.

Great care should be taken to prevent air bubbles in castings; and the more time there can be allowed for cooling the better, because the iron will be tougher than when rapidly cooled; slow cooling answers the same purpose as annealing.

In making patterns for cast iron, an allowance of about one-eighth of an inch per foot must be made for the contraction of the metal in cooling. Also the patterns that require it should be slightly bevelled to allow of their being drawn out of the sand without injuring the impression; about one-sixteenth of an inch in six inches is sufficient for this purpose.

In notes at the foot of each Table, the mode of applying these Tables to other materials is shown, which will be useful in exhibiting the comparative strength of different bodies when applied to the same purpose, as well as in giving the proportions of these materials for supporting a given load.

TABLE I.—Art. 5. *A Table of the Depths of Square Beams or Bars cwt. to 500 tons, when supported at the ends, and loaded in the middle;*

Lengths in feet.		4	6	8	10	12	14	16	18	20
Weight in tons.	Weight in lbs.	Depth inches.	Depth inches.	Depth inches.	Depth inches.	Depth inches.	Depth inches.	Depth inches.	Depth inches.	Depth inches.
$\frac{1}{20}$	112	1·2	1·4	1·7	1·9	2·0	2·2	2·4	2·5	2·6
$\frac{1}{10}$	224	1·4	1·7	2·0	2·2	2·4	2·6	2·8	3·0	3·1
$\frac{3}{20}$	336	1·6	1·9	2·2	2·4	2·7	2·9	3·1	3·3	3·4
$\frac{1}{5}$	448	1·7	2·0	2·4	2·6	2·9	3·1	3·3	3·5	3·7
$\frac{1}{4}$	560	1·8	2·2	2·5	2·8	3·0	3·3	3·5	3·7	3·9
$\frac{3}{10}$	672	1·8	2·2	2·6	2·9	3·2	3·4	3·7	3·9	4·1
$\frac{7}{20}$	784	1·9	2·3	2·7	3·0	3·3	3·6	3·8	4·1	4·2
$\frac{2}{5}$	896	2·0	2·4	2·8	3·1	3·4	3·7	3·9	4·2	4·4
$\frac{9}{20}$	1008	2·0	2·5	2·9	3·2	3·5	3·8	4·0	4·3	4·5
$\frac{1}{2}$	1120	2·1	2·6	3·0	3·3	3·6	3·9	4·2	4·4	4·7
$\frac{11}{20}$	1232	2·1	2·6	3·0	3·4	3·7	4·0	4·3	4·5	4·8
$\frac{3}{5}$	1344	2·2	2·7	3·1	3·5	3·8	4·1	4·4	4·7	4·9
$\frac{13}{20}$	1456	2·2	2·7	3·1	3·5	3·8	4·2	4·4	4·7	4·9
$\frac{7}{10}$	1568	2·3	2·8	3·2	3·6	3·9	4·2	4·5	4·8	5·0
$\frac{3}{4}$	1680	2·3	2·8	3·2	3·6	4·0	4·3	4·6	4·9	5·2
$\frac{4}{5}$	1792	2·4	2·9	3·3	3·7	4·0	4·4	4·7	5·0	5·2
$\frac{17}{20}$	1904	2·4	2·9	3·4	3·8	4·1	4·4	4·7	5·0	5·3
$\frac{9}{10}$	2016	2·4	3·0	3·4	3·8	4·2	4·5	4·8	5·1	5·4
$\frac{19}{20}$	2128	2·5	3·0	3·5	3·9	4·2	4·6	4·9	5·2	5·4
1	2240	2·5	3·0	3·5	3·9	4·3	4·6	4·9	5·2	5·5
$1\frac{1}{4}$	2800	2·6	3·2	3·7	4·1	4·5	4·9	5·2	5·5	5·8
Deflex. in inches.		·1	·15	·2	·25	·3	·35	·4	·45	·5

[11] The Table was calculated by the rule in art. 257.

of Cast Iron, of different lengths, to sustain weights[12] *of from one the deflexion not to exceed $\frac{1}{40}$ of an inch for each foot in length.*[11]

22	24	26	28	30	32	34	36	38	40	
Depth inches.	Depth inches.	Depth inches.	Depth inches.	Depth inches.	Depth inches.	Depth inches.	Depth inches.	Depth inches.	Depth inches.	Weight.
2·7	2·9	3·0	3·1	3·2	3·3	3·4	3·5	3·6	3·7	1 cwt.
3·3	3·4	3·6	3·7	3·8	3·9	4·1	4·2	4·3	4·4	2 —
3·6	3·8	3·9	4·1	4·2	4·3	4·5	4·6	4·7	4·8	3 —
3·9	4·0	4·2	4·3	4·5	4·7	4·8	4·9	5·0	5·2	4 —
4·1	4·3	4·4	4·6	4·8	4·9	5·1	5·2	5·4	5·5	5 —
4·3	4·5	4·6	4·8	5·0	5·1	5·3	5·4	5·6	5·8	6 —
4·4	4·6	4·8	5·0	5·2	5·4	5·5	5·7	5·9	6·0	7 —
4·6	4·8	5·0	5·2	5·4	5·6	5·7	5·9	6·0	6·2	8 —
4·7	4·9	5·1	5·3	5·5	5·7	5·9	6·0	6·2	6·4	9 —
4·9	5·2	5·3	5·4	5·7	5·9	6·0	6·2	6·4	6·5	10 —
5·0	5·3	5·4	5·6	5·8	6·0	6·2	6·4	6·5	6·7	11 —
5·1	5·3	5·5	5·7	5·9	6·1	6·3	6·5	6·7	6·8	12 —
5·2	5·4	5·6	5·9	6·0	6·2	6·5	6·6	6·8	7·0	13 —
5·3	5·5	5·7	6·0	6·1	6·4	6·6	6·7	6·9	7·1	14 —
5·4	5·6	5·8	6·1	6·2	6·5	6·7	6·8	7·0	7·2	15 —
5·5	5·7	5·9	6·2	6·4	6·6	6·8	6·9	7·2	7·4	16 —
5·5	5·8	6·0	6·2	6·5	6·7	6·9	7·1	7·3	7·5	17 —
5·6	5·9	6·1	6·4	6·6	6·8	7·0	7·2	7·4	7·6	18 —
5·7	6·0	6·2	6·5	6·7	6·9	7·1	7·3	7·5	7·7	19 —
5·8	6·0	6·3	6·5	6·8	7·0	7·2	7·4	7·5	7·8	1 ton.
6·1	6·4	6·6	6·9	7·2	7·4	7·6	7·8	8·0	8·2	$1\frac{1}{4}$ —
·55	·6	·65	·7	·75	·8	·85	·9	·95	1·0	Defl. in.

[12] The weight of the load to be supported must include the weight of the beam. To find the weight of a beam, multiply the area of the section in inches by the length in feet and by 3·2, which will give the weight in lbs.

TABLE I.—*Of the Stiffness*

Lengths in feet.		4	6	8	10	12	14	16	18	20
Weight in tons.	Weight in lbs.	Depth inches.	Depth inches.	Depth inches.	Depth inches.	Depth inches.	Depth inches.	Depth inches.	Depth inches.	Depth inches.
1½	3,360	2·8	3·4	3·9	4·3	4·7	5·1	5·5	5·8	6·1
1¾	3,920	2·9	3·5	4·0	4·5	4·9	5·3	5·7	6·0	6·3
2	4,480	2·9	3·5	4·1	4·7	5·1	5·5	5·9	6·2	6·5
2½	5,600	3·1	3·8	4·4	4·9	5·5	5·8	6·2	6·6	6·9
3	6,720	3·3	4·0	4·6	5·1	5·7	6·1	6·5	6·9	7·3
3½	7,840	3·4	4·1	4·8	5·3	5·8	6·3	6·7	7·1	7·5
4	8,960	3·5	4·3	4·9	5·5	6·0	6·5	7·0	7·4	7·8
4½	10,080		4·4	5·1	5·7	6·2	6·7	7·2	7·6	8·0
5	11,200		4·5	5·2	5·8	6·4	6·9	7·4	7·8	8·2
6	13,440			5·5	6·1	6·7	7·2	7·7	8·2	8·6
7	15,680			5·7	6·3	6·9	7·5	8·0	8·5	8·9
8	17,920			5·9	6·6	7·2	7·8	8·3	8·8	9·3
9	20,160			6·0	6·8	7·4	8·0	8·5	9·0	9·5
10	22,400				6·9	7·6	8·2	8·8	9·3	9·8
11	24,640				7·1	7·8	8·4	9·0	9·5	10·0
12	26,880				7·2	7·9	8·6	9·2	9·7	10·2
13	29,120				7·4	8·1	8·8	9·4	9·9	10·4
14	31,360				7·5	8·3	8·9	9·5	10·1	10·6
15	33,600				7·7	8·4	9·1	9·7	10·3	10·8
16	35,840				7·8	8·5	9·2	9·8	10·4	11·0
17	38,080				7·9	8·7	9·4	10·0	10·6	11·2
18	40,320				8·0	8·8	9·5	10·1	10·8	11·3
19	42,560				8·1	8·9	9·6	10·3	10·9	11·5
Deflex. in inches.		·1	·15	·2	·25	·3	·35	·4	·45	·5

If the depth of a cast iron bar be multiplied by 0·937, the product will be the depth of a square bar of wrought iron of equal stiffness.

of Beams (*continued*).

22	24	26	28	30	32	34	36	38	40	
Depth inches.	Depth inches.	Depth inches.	Depth inches.	Depth inches.	Depth inches.	Depth inches.	Depth inches.	Depth inches.	Depth inches.	Weight in tons.
6·4	6·7	7·0	7·2	7·5	7·7	8·0	8·2	8·4	8·6	$1\frac{1}{2}$
6·7	6·9	7·2	7·5	7·7	8·0	8·2	8·5	8·7	8·9	$1\frac{3}{4}$
6·8	7·2	7·6	7·7	8·0	8·3	8·5	8·7	9·0	9·2	2
7·3	7·6	7·9	8·2	8·5	8·8	9·0	9·3	9·6	9·8	$2\frac{1}{2}$
7·6	7·9	8·3	8·6	8·9	9·2	9·4	9·7	10·0	10·1	3
7·9	8·2	8·6	8·9	9·2	9·5	9·8	10·1	10·4	10·6	$3\frac{1}{2}$
8·2	8·5	8·9	9·2	9·5	9·8	10·1	10·4	10·7	11·0	4
8·4	8·8	9·1	9·5	9·8	10·1	10·4	10·8	11·0	11·4	$4\frac{1}{2}$
8·6	9·0	9·4	9·7	10·1	10·4	10·7	11·0	11·2	11·6	5
9·0	9·4	9·8	10·2	10·5	10·9	11·2	11·5	11·9	12·1	6
9·4	9·8	10·2	10·6	11·0	11·3	11·7	12·0	12·3	12·7	7
9·7	10·1	10·6	10·9	11·3	11·7	12·0	12·4	12·8	13·1	8
10·0	10·4	10·9	11·3	11·7	12·0	12·4	12·8	13·1	13·5	9
10·3	10·7	11·2	11·6	12·0	12·4	12·8	13·1	13·5	13·8	10
10·5	11·0	11·5	11·9	12·3	12·7	13·1	13·5	13·8	14·2	11
10·8	11·2	11·7	12·1	12·5	13·0	13·4	13·7	14·1	14·5	12
11·0	11·5	11·9	12·4	12·8	13·2	13·6	14·0	14·4	14·7	13
11·1	11·7	12·1	12·6	13·0	13·4	13·8	14·2	14·6	15·0	14
11·4	11·9	12·3	12·8	13·2	13·7	14·1	14·5	14·9	15·3	15
11·5	12·0	12·5	13·0	13·5	13·9	14·3	14·7	15·1	15·5	16
11·7	12·2	12·7	13·2	13·7	14·1	14·5	14·9	15·4	15·8	17
11·9	12·4	12·9	13·4	13·9	14·3	14·7	15·1	15·6	16·0	18
12·0	12·6	13·1	13·6	14·1	14·5	15·0	15·4	15·8	16·2	19
·55	·6	·65	·7	·75	·8	·85	·9	·95	1·0	Deflex.

If the depth of a cast iron beam be multiplied by 1·83, the product will be the depth of a square beam of oak of equal stiffness.

TABLE I.—*Of the Stiffness*

Lengths in feet.		4	6	8	10	12	14	16	18	20
Weight in tons.	Weight in lbs.	Depth inches.	Depth inches.	Depth inches.	Depth inches.	Depth inches.	Depth inches.	Depth inches.	Depth inches.	Depth inches.
20	44,800					9·0	9·7	10·4	11·0	11·6
22	49,280					9·2	10·0	10·7	11·3	11·9
24	53,760					9·4	10·2	10·9	11·5	12·2
26	58,240					9·6	10·4	11·1	11·8	12·4
28	62,720					9·8	10·6	11·4	12·0	12·7
30	67,200						10·8	11·5	12·2	12·9
32	71,680						11·0	11·7	12·4	13·1
34	76,160						11·1	11·9	12·6	13·3
36	80,640						11·3	12·0	12·8	13·4
38	85,120						11·4	12·2	13·0	13·6
40	89,600							12·4	13·1	13·8
42	94,080							12·5	13·3	14·0
44	98,560							12·7	13·5	14·2
46	103,040							12·8	13·6	14·3
48	107,520							13·0	13·7	14·5
50	112,000								13·9	14·6
52	116,480								14·0	14·7
54	120,960								14·1	14·9
56	125,440								14·3	15·0
58	129,920								14·4	15·1
60	134,400								14·5	15·3
65	145,600								14·8	15·6
70	156,800								15·1	15·9
Deflex. in inches.		·1	·15	·2	·25	·3	·35	·4	·45	·5

The depth of a yellow fir beam of equal stiffness may be found by multiplying the depth of the cast iron one by 1·71.

of Beams (continued).

22	24	26	28	30	32	34	36	38	40	
Depth inches.	Depth inches.	Depth inches.	Depth inches.	Depth inches.	Depth inches.	Depth inches.	Depth inches.	Depth inches.	Depth inches.	Weight in tons.
12·2	12·7	13·2	13·8	14·2	14·7	15·1	15·6	16·0	16·4	20
12·5	13·0	13·6	14·1	14·6	15·1	15·5	15·9	16·4	16·8	22
12·8	13·4	13·9	14·4	14·9	15·4	15·9	16·3	16·8	17·2	24
13·0	13·6	14·2	14·7	15·2	15·7	16·2	16·7	17·1	17·6	26
13·3	13·9	14·4	15·0	15·5	16·0	16·5	17·0	17·4	17·9	28
13·5	14·1	14·7	15·2	15·7	16·3	16·8	17·3	17·7	18·2	30
13·7	14·3	14·9	15·5	16·0	16·5	17·0	17·5	18·0	18·5	32
13·9	14·5	15·1	15·7	16·2	16·8	17·3	17·8	18·3	18·8	34
14·1	14·7	15·3	15·9	16·5	17·0	17·5	18·0	18·5	19·0	36
14·3	14·9	15·5	16·1	16·7	17·2	17·8	18·3	18·8	19·3	38
14·5	15·1	15·7	16·4	16·9	17·5	18·0	18·5	19·1	19·5	40
14·7	15·3	15·9	16·5	17·1	17·7	18·2	18·7	19·3	19·8	42
14·9	15·5	16·1	16·8	17·4	17·9	18·5	19·0	19·5	20·0	44
15·0	15·7	16·3	17·0	17·6	18·1	18·7	19·2	19·8	20·3	46
15·2	15·9	16·5	17·1	17·7	18·3	18·8	19·4	20·0	20·5	48
15·3	16·0	16·6	17·3	17·9	18·5	19·0	19·6	20·1	20·7	50
15·5	16·2	16·8	17·5	18·1	18·7	19·2	19·8	20·3	21·0	52
15·6	16·3	17·0	17·6	18·2	18·8	19·4	19·9	20·5	21·1	54
15·8	16·5	17·1	17·8	18·4	19·0	19·6	20·1	20·7	21·3	56
15·9	16·6	17·3	17·9	18·5	19·2	19·7	20·3	20·9	21·4	58
16·0	16·7	17·4	18·1	18·7	19·3	19·9	20·5	21·1	21·6	60
16·4	17·1	17·8	18·5	19·1	19·8	20·4	20·9	21·5	22·1	65
16·7	17·4	18·2	18·8	19·5	20·1	20·8	21·3	22·0	22·5	70
·55	·60	·65	·70	·75	·80	·85	·90	·95	1·0	Deflex.

TABLE I.—*Of the Stiffness*

Lengths in feet.		4	6	8	10	12	14	16	18	20
Weight in tons.	Weight in pounds.	Depth inches.	Depth inches.	Depth inches.	Depth inches.	Depth inches.	Depth inches.	Depth inches.	Depth inches.	Depth inches.
75	168,000									
80	179,200									
85	190,400									
90	201,600									
95	212,800									
100	224,000									
110	246,400									
120	268,800									
130	291,200									
140	313,600									
150	336,000									
160	358,400									
170	380,800									
180	403,200									
190	425,600									
200	448,000									
250	560,000									
300	672,000									
350	784,000									
400	896,000									
450	1,008,000									
500	1,120,000									
Deflex. in inches.		·10	·15	·20	·25	·30	·35	·40	·45	·50

of Beams (continued).

22	24	26	28	30	32	34	36	38	40	
Depth inches.	Depth inches.	Depth inches.	Depth inches.	Depth inches.	Depth inches.	Depth inches.	Depth inches.	Depth inches.	Depth inches.	Weight in tons.
17·0	17·7	18·5	19·2	19·8	20·5	21·3	21·7	22·3	22·9	75
17·2	18·0	18·7	19·4	20·1	20·7	21·4	22·0	22·6	23·2	80
17·5	18·3	19·0	19·7	20·4	21·0	21·7	22·4	23·0	23·6	85
17·8	18·6	19·3	20·0	20·7	21·4	22·1	22·7	23·3	23·9	90
18·0	18·8	19·5	20·3	21·0	21·7	22·4	23·0	23·6	24·3	95
	19·0	19·8	20·6	21·3	22·0	22·6	23·3	23·9	24·5	100
		20·3	21·0	21·8	22·5	23·2	23·8	24·5	25·1	110
		20·8	21·5	22·3	23·0	23·7	24·4	25·0	25·7	120
			22·0	22·7	23·5	24·2	24·9	25·5	26·3	130
				23·2	23·9	24·6	25·4	26·0	26·7	140
				23·6	24·3	25·0	25·8	26·5	27·2	150
				23·9	24·7	25·5	26·2	26·9	27·6	160
					25·1	25·9	26·6	27·4	28·1	170
					25·5	26·3	27·0	27·8	28·5	180
					25·9	26·7	27·4	28·2	28·9	190
					26·2	27·0	27·7	28·5	29·2	200
					27·6	28·5	29·4	30·1	31·0	250
					28·9	29·9	30·7	31·5	32·0	300
					30·0	31·0	32·0	33·0	33·7	350
					31·1	32·0	32·9	33·9	34·7	400
					32·0	33·1	34·0	34·8	35·8	450
						33·8	34·8	35·7	36·7	500
·55	·60	·65	·70	·75	·80	·85	·90	·95	1·0	Deflex.

TABLE II.—ART. 6. *A Table showing the weight or pressure a its elastic force, when it is supported at the ends, and loaded that weight will produce. This Table was calculated by the*

Lengths	1 foot.		2 feet.		3 feet.		4 feet.		5 feet.	
Depths.	Weight in lbs.	Defl. in inches.	Weight in lbs.	Defl. in inches.	Weight in lbs.	Defl. in inches.	Weight in lbs.	Defl. in inches.	Weight in lbs.	Defl. in inches.
1 in.	850	·02	425	·08	283	·18	212	·32	170	·5
1½ —	1,912	·014	956	·053	637	·12	477	·21	383	·33
2 —			1,700	·04	1,132	·09	848	·16	680	·25
2½ —			2,656	·032	1,769	·072	1,325	·128	1,062	·2
3 —					2,547	·06	1,908	·11	1,530	·167
3½ —					3,467	·052	2,597	·092	2,082	·143
4 —							3,392	·08	2,720	·125
4½ —							4,293	·071	3,442	·111
5 —									4,250	·1
6 —									6,120	·083
7 —										
8 —										
9 —										
10 —										
11 —										
12 —										
13 —										
14 —										
Fixed at one end.	¼ foot.		½ foot.		¾ foot.		1 foot.		1¼ foot.	

REMARK.—The load shown by this Table is the greatest a beam should ever sustain, and therefore, in calculating this load, ample allowance must be made for accidents, and the weight of the beam itself must be included. The weight of the beam may be easily calculated by the Rule in the note to Table I.

beam of cast iron, one inch in breadth, will sustain without destroying in the middle of its length, and also the deflexion in the middle which Equation, art. 143.[13]

6 feet.		7 feet.		8 feet.		9 feet.		10 feet.		Lengths
Weight in lbs.	Defl. in inches.	Weight in lbs.	Defl. in inches.	Weight in lbs.	Defl. in inches.	Weight in lbs.	Defl. in inches.	Weight in lbs.	Defl. in inches.	Depths.
142	·72	121	·98	106	1·28	95	1·62	85	2·0	1 in.
320	·48	273	·65	239	·85	214	1·08	192	1·34	1½—
568	·36	484	·49	425	·64	380	·81	340	1·0	2 —
887	·29	756	·39	662	·51	594	·65	531	·8	2½—
1,278	·24	1,089	·33	954	·426	855	·54	765	·66	3 —
1,739	·205	1,482	·28	1,298	·365	1,164	·46	1,041	·57	3½—
2,272	·18	1,936	·245	1,700	·32	1,520	·405	1,360	·5	4 —
2,875	·16	2,450	·217	2,146	·284	1,924	·36	1,721	·443	4½—
3,560	·144	3,050	·196	2,650	·256	2,375	·32	2,125	·4	5 —
5,112	·12	4,356	·163	3,816	·213	3,420	·27	3,060	·33	6 —
6,958	·103	5,929	·14	5,194	·183	4,655	·23	4,165	·29	7 —
9,088	·09	7,744	·123	6,784	·16	6,080	·203	5,440	·25	8 —
		9,801	·109	8,586	·142	7,695	·18	6,885	·22	9 —
		12,100	·098	10,600	·128	9,500	·162	8,500	·2	10 —
				12,826	·117	11,495	·15	10,285	·182	11 —
				15,264	·107	13,680	·135	12,240	·17	12 —
						16,100	·125	14,400	·154	13 —
						18,600	·115	16,700	·143	14 —
1½ foot.		1¾ foot.		2 feet.		2¼ feet.		2½ feet.		

[13] A piece increased to n times the depth has its strength increased $3\,(n-1)+\frac{1}{n}$ times.

TABLE II.—*Of the Strength*

Lengths	12 feet.		14 feet.		16 feet.		18 feet.		20 feet.	
Depths.	Weight in lbs.	Defl. in inches.	Weight in lbs.	Defl. in inches.	Weight in lbs.	Defl. in inches.	Weight in lbs.	Defl. in inches.	Weight in lbs.	Defl. in inches.
2 in.	283	1·44	243	1·96	212	2·56	189	3·24	170	4·0
3 —	637	·96	546	1·31	478	1·71	425	2·16	382	2·67
4 —	1,133	·72	971	·98	849	1·28	755	1·62	680	2·08
5 —	1,771	·58	1,518	·78	1,328	1·02	1,180	1·29	1,062	1·6
6 —	2,548	·48	2,184	·65	1,912	·85	1,699	1·08	1,530	1·34
7 —	3,471	·41	2,975	·58	2,603	·73	2,314	·93	2,082	1·14
8 —	4,532	·36	3,884	·49	3,396	·64	3,020	·81	2,720	1·00
9 —	5,733	·32	4,914	·44	4,302	·57	3,825	·72	3,438	·89
10 —	7,083	·288	6,071	·392	5,312	·512	4,722	·648	4,250	·8
11 —	8,570	·26	7,346	·36	6,428	·47	5,714	·59	5,142	·73
12 —	10,192	·24	8,736	·33	7,648	·43	6,796	·54	6,120	·67
13 —	11,971	·22	10,260	·307	8,978	·39	7,980	·49	7,182	·61
14 —	13,883	·21	11,900	·28	10,412	·36	9,255	·46	8,330	·57
15 —	15,937	·19	13,660	·26	11,952	·34	10,624	·43	9,562	·533
16 —	18,128	·18	15,536	·245	13,584	·32	12,080	·403	10,880	·5
17 —	20,500	·17	17,500	·23	15,353	·3	13,647	·38	12,282	·47
18 —	22,932	·16	19,656	·217	17,208	·284	15,700	·36	13,752	·442
19 —	25,404	·152	21,800	·207	19,053	·27	16,935	·34	15,242	·42
20 —	28,332	·144	24,284	·195	21,248	·256	18,888	·324	17,000	·4
Fixed at one end.	3 feet.		3½ feet.		4 feet.		4½ feet.		5 feet.	

of Beams (continued).

22 feet.		24 feet.		26 feet.		28 feet.		30 feet.		Lengths
Weight in lbs.	Defl. in inches.	Weight in lbs.	Defl. in inches.	Weight in lbs.	Defl. in inches.	Weight in lbs.	Defl. in inches.	Weight in lbs.	Defl. in inches.	Depths.
154	4·84	142	5·76	131	6·76	121	7·84	113	9·0	2 in.
347	3·23	318	3·84	294	4·51	273	5·23	255	6·0	3—
618	2·42	566	2·88	523	3·38	485	3·92	453	4·5	4—
966	1·93	885	2·30	817	2·70	759	3·14	708	3·6	5—
1,390	1·61	1,274	1·92	1,176	2·25	1,092	2·61	1,019	3·0	6—
1,893	1·38	1,735	1·65	1,602	1·93	1,487	2·24	1,388	2·57	7—
2,472	1·21	2,264	1·44	2,092	1·69	1,940	1·96	1,812	2·25	8—
3,123	1·07	2,862	1·28	2,646	1·50	2,457	1·74	2,295	2·0	9—
3,863	·968	3,541	1·152	3,269	1·352	3,035	1·568	2,833	1·8	10—
4,675	·88	4,285	1·05	3,955	1·23	3,673	1·425	3,428	1·64	11—
5,560	·81	5,096	·96	4,704	1·13	4,368	1·31	4,076	1·5	12—
6,529	·74	5,985	·886	5,525	1·04	5,130	1·21	4,788	1·38	13—
7,573	·69	6,941	·824	6,408	·965	5,950	1·12	5,553	1·28	14—
8,692	·645	7,967	·75	7,355	·9	6,829	1·03	6,374	1·2	15—
9,888	·63	9,056	·72	8,368	·84	7,760	·98	7,248	1·13	16—
11,166	·567	10,235	·673	9,447	·79	8,773	·92	8,188	1·06	17—
12,492	·54	11,448	·64	10,584	·75	9,828	·87	9,180	1·0	18—
13,857	·51	12,702	·607	11,725	·71	10,887	·825	10,161	·95	19—
15,452	·484	14,164	·576	13,076	·676	12,140	·784	11,332	·9	20—
5½ feet.		6 feet.		6½ feet.		7 feet.		7½ feet.		

TABLE II.—*Of the Strength*

Lengths	12 feet.		14 feet.		16 feet.		18 feet.		20 feet.	
Depths.	Weight in lbs.	Defl. in inches.	Weight in lbs.	Defl. in inches.	Weight in lbs.	Defl. in inches.	Weight in lbs.	Defl. in inches.	Weight in lbs.	Defl. in inches.
21 in.	31,230	·138	26,770	·186	23,428	·245	20,825	·31	18,742	·382
22—	34,500	·131	29,300	·178	25,712	·235	22,855	·295	20,570	·365
23—	37,600	·127	32,000	·17	28,103	·225	24,980	·282	22,482	·35
24—	40,768	·12	34,944	·163	30,592	·216	27,184	·27	24,480	·335
25—			37,700	·156	33,203	·21	29,514	·26	26,562	·32
26—			40,900	·15	35,912	·197	31,922	·25	28,730	·307
27—			44,000	·143	38,728	·19	34,425	·24	30,982	·297
28—			47,300	·14	41,650	·183	37,022	·23	33,320	·286
29—					44,678	·176	39,7,14	·223	35,742	·275
30—					47,808	·170	42,498	·216	38,250	·266
31—					51,053	·164	45,380	·207	40,842	·257
32—					54,400	·16	48,371	·202	43,520	·25
33—							51,425	·196	46,282	·242
34—							54,586	·19	49,130	·235
35—							57,847	·185	52,062	·228
36—							61,200	·18	55,080	·222
Fixed at one end.	3 feet.		$3\frac{1}{2}$ feet.		4 feet.		$4\frac{1}{2}$ feet.		5 feet.	

If the weight a cast iron bar will support be multiplied by 1·12, the product will be the weight a WROUGHT IRON bar of the same size will support. And the flexure of the wrought iron bar will be found by multiplying the flexure of the cast iron one by 0·86.

The strength of good OAK is one-fourth of the strength of cast iron; therefore an oak beam will bear one-fourth of the load of a cast iron one of the same size. And the flexure of the oak beam will be found by multiplying the flexure of the cast iron one by 2·8.

of Beams (continued).

22 feet.		24 feet.		26 feet.		28 feet.		30 feet.		Lengths
Weight in lbs.	Defl. in inches.	Weight in lbs.	Defl. in inches.	Weight in lbs.	Defl. in inches.	Weight in lbs.	Defl. in inches.	Weight in lbs.	Defl. in inches.	Depths.
17,036	·45	15,618	·55	14,417	·645	13,387	·75	12,495	·86	21 in.
18,700	·44	17,141	·525	15,823	·615	14,693	·71	13,713	·815	22—
20,439	·42	18,735	·5	17,286	·59	16,059	·68	14,988	·78	23—
22,240	·402	20,384	·48	18,816	·565	17,492	·665	16,304	·75	24—
24,148	·387	22,135	·46	20,432	·54	18,973	·625	17,708	·72	25—
26,118	·375	23,941	·443	22,100	·52	20,521	·607	19,153	·695	26—
28,166	·36	25,819	·427	23,832	·5	22,130	·58	20,655	·667	27—
30,290	·347	27,766	·41	25,630	·48	23,800	·56	22,213	·645	28—
32,493	·333	29,785	·395	27,494	·462	25,530	·54	23,828	·62	29—
34,767	·322	31,869	·384	29,421	·450	27,315	·522	25,497	·60	30—
37,148	·31	34,035	·37	31,417	·435	29,173	·505	27,228	·58	31—
39,563	·302	36,266	·36	33,477	·42	31,086	·49	29,013	·56	32—
42,075	·293	38,568	·35	35,602	·41	33,058	·47	30,855	·545	33—
44,663	·283	40,941	·336	37,792	·395	35,093	·46	32,753	·53	34—
47,329	·276	43,385	·329	40,048	·386	37,187	·448	34,708	·514	35—
50,073	·269	45,900	·32	42,369	·375	39,343	·435	36,720	·5	36—
5½ feet.		6 feet.		6½ feet.		7 feet.		7½ feet.		

To find the weight a beam of YELLOW FIR will bear, multiply the weight a cast iron one of the same size will bear by 0·3. And to find its flexure, multiply the flexure of a cast iron one by 2·6.

In the same manner the Table may be applied to find the strength of any other material of which the proportional strength in respect to cast iron is known. See the alphabetical Table at the end of this work.

TABLE III.[14]—ART. 7. *A Table to show the weight or pressure a cylindrical pillar or column of cast iron will sustain, with safety, in hundred weights.*

Length or height.	2 ft.	4 ft.	6 ft.	8 ft.	10ft.	12ft.	14ft.	16ft.	18ft.	20ft.	22ft.	24ft.	
Dia-meter.	Wt. in cwts.	Wt. in cwts.	Wt. in cwts.	Wt. in cwts.	Wt. in cwts.	Wt. in cwts.	Wt. in cwts.	Wt. in cwts.	Wt. in cwts.	Wt. in cwts.	Wt. in cwts.	Wt. in cwts.	Dia-meter.
1 in.	18	12	8	5	3	2	2	1	1	1			1 in.
1½—	44	36	28	19	16	12	9	7	6	5	4	3	1½—
2 —	82	72	60	49	40	32	26	22	18	15	13	11	2 —
2½—	129	119	105	91	77	65	55	47	40	34	29	25	2½—
3 —	188	178	163	145	128	111	97	84	73	64	56	49	3 —
3½—	257	247	232	214	191	172	156	135	119	106	94	83	3½—
4 —	337	326	310	288	266	242	220	198	178	160	144	130	4 —
4½—	429	418	400	379	354	327	301	275	251	229	208	189	4½—
5 —	530	522	501	479	452	427	394	365	337	310	285	262	5 —
6 —	616	607	592	573	550	525	497	469	440	413	386	360	6 —
7 —	1040	1032	1013	989	959	924	887	848	808	765	725	686	7 —
8 —	1344	1333	1315	1289	1259	1224	1185	1142	1097	1052	1005	959	8 —
9 —	1727	1716	1697	1672	1640	1603	1561	1515	1467	1416	1364	1311	9 —
10 —	2133	2119	2100	2077	2045	2007	1964	1916	1865	1811	1755	1697	10 —
11 —	2580	2570	2550	2520	2490	2450	2410	2358	2305	2248	2189	2127	11 —
12 —	3074	3050	3040	3020	2970	2930	2900	2830	2780	2730	2670	2600	12 —

This Table was calculated by Equation xviii. art. 290. It is one of those cases where a Table is most useful even to the quickest calculator, because the weight to be supported and the length being given, a quadratic equation must be solved to find the diameter: here it is found by inspection. This Table does not admit of accurate application to other materials, on account of the form of the equation. It will be nearly correct for wrought iron, but is not applicable to timber.

[14] This Table has no solid basis. The very ingenious reasoning, from which the formula is deduced by which the Table was calculated, depends upon assumptions which Mr. Tredgold was induced to adopt through want of experimental data. See Mr. Barlow's Report on the Strength of Materials, 2nd vol. of the British Association. An abstract of an experimental research, to supply this deficiency, will be given in the "Additions."—EDITOR.

SECTION II.

EXPLANATION OF THE TABLES, WITH EXAMPLES OF THEIR USE.

EXPLANATION OF THE FIRST TABLE.

8. The first Table (page 12, art. 5,) shows by inspection, the dimensions of square beams to sustain weights or pressures of from one hundred weight to 500 tons; so as not to be bent or deflected in the middle, more than one-fortieth of an inch for each foot in length.

The length is the distance between the supports, as A B, fig. 1, Plate I., and the stress, whether it be from weight or pressure, is supposed to act at the middle of the length, as at C in the figure. The breadth and depth are supposed to be the same in every part of the length, and equal to one another.

The horizontal row of figures at the top of the Table contains the lengths in feet.

The columns, at the outsides, contain the weights in cwts. and tons, and the second column, on the left-hand side, contains the weights in pounds avoirdupois.

The horizontal row of figures at the bottom shows the deflexion for each length. The other columns show the depths in inches.

EXPLANATION OF THE SECOND TABLE.

9. The second Table (page 20, art. 6,) is intended to show the greatest weight a beam of cast-iron will bear in the middle of its length, when it is loaded with as much as it will bear, so as to recover its natural form when the load is removed. If a beam be loaded beyond that point, the equilibrium of its parts is destroyed, and it takes a permanent set. Also, in a beam so loaded beyond its strength, the deflexion becomes irregular, increasing very rapidly in proportion to the load.

The horizontal row of figures along the top of the Table contains the lengths in feet, that is, the distances between the points of support; and the horizontal row at bottom, the length of a beam supported or fixed at one end only, which with the same load would have the same deflexion.

The columns on the outsides contain the depths in inches.

The other columns contain the weights in pounds avoirdupois, and the deflexions they would produce in inches and decimal parts, when the beams will be only just capable of restoring themselves.

The breadth of each beam is one inch, therefore the Table shows the utmost weight a beam of one inch in breadth should have to bear; and a piece five inches in breadth will bear five times as much, and so of any other breadth.

EXPLANATION OF THE THIRD TABLE.[1]

10. The third Table (page 26, art. 7,) shows by inspection the weight or pressure a cylindrical pillar or column of cast iron will bear with safety. The pressure is expressed in cwts. and is computed on the supposition that the pillar is under the most unfavourable circumstances for resisting the stress, which happens, when, from settlements, imperfect fitting, or other causes, the direction of the stress is in the surface of the pillar, as shown in fig. 31, Plate IV.

The horizontal row of figures along the top of the Table contains the lengths or heights of the pillars in feet.

The outside vertical columns of the Table contain the diameters of the pillars in inches.

The other vertical columns of the Table show the weight in cwts. which a cast iron pillar, of the height at the top of the column, and of the diameter in the side columns, will support with safety. Consequently, of the height, the diameter, and the

---

[1] See note to that Table.—EDITOR.

weight to be supported, any two being given, the other will be found by inspection.

EXAMPLES AND USE OF THE TABLES.

11. *Example* 1. To find the depth of a square bar of cast iron, twenty feet in length, that would support ten tons, the deflexion not exceeding half an inch.

Find the column in Table I. which has the length twenty feet at the top, and in that column, and opposite to ten tons in either of the side columns, will be found the proper depth for the bar, which is 9·8 inches.

If the depth 9·8 be multiplied by 1·71, it will give the depth of a square beam of fir that would support the same load with the same deflexion. Thus, $1·71 \times 9·8 = 16·76$ inches nearly, the depth of the fir beam.

If the depth of an oak beam be required, multiply by 1·83; thus $1·83 \times 9·8 = 17·93$ inches, the depth of an oak beam.

12. *Example* 2. Required the weight a cast iron beam would support without impairing its elastic force, the length, breadth, and depth being given?

Let the length be twenty feet, and the breadth the same as the depth, ten inches. In the second Table, under the length twenty feet, and opposite the depth ten inches, we find the weight 4,250 ℔s. for the load a beam one inch in breadth would bear;

and this multiplied by 10, gives 42,500 ℔s., or nearly nineteen tons; and the deflexion would be 0·8 inches, but the weight of the beam itself would be nearly three tons, and its effect the same as if half the three tons were applied in the middle; consequently the greatest load that the beam should be liable to sustain should not exceed seventeen tons and a half.

An oak beam of the same size would support only one-fourth of 42,500 ℔s. or 10,625 ℔s.; and its deflexion in the middle would be 0·8 multiplied by 2·8=2·24 inches.

A fir beam of the same size would support three-tenths of 42,500 ℔s. = 12,750 ℔s.; and its deflexion in the middle would be $0{\cdot}8 \times 2{\cdot}6 = 2{\cdot}08$ inches.

A wrought iron bar of the same size would support 1·12 times the weight of the cast iron one, that is, $42{,}500 \times 1{\cdot}12 = 47{,}600$ ℔s.; and its deflexion in the middle would be 0·8 multiplied by 0·86 =0·688 inches. But the reader will remember that wrought iron possesses this great stiffness only in consequence of the operations of forging or rolling, and these operations have very little effect where the thickness is considerable.

13. There are cases where a greater degree of flexure may be allowed, and there are others where it ought to be less; but I consider that to which the first Table is calculated as nearly the mean, and it is easy to make any variation in this respect.

Example 3. Let it be required to find the depth of a square cast iron bar to support ten tons without more deflexion than one-tenth of an inch, the length being twenty feet.

By examining the deflexion for twenty feet at the foot of the column in Table I. it will be found five times one-tenth of an inch; hence take the depth opposite five times the weight or fifty tons, which is 14·6 inches, the depth required.

14. *Example* 4. Find the depth of a square bar of cast iron to support ten tons, the deflexion not to exceed one inch, the length being twenty feet.

This degree of deflexion is double that at the foot of the column headed 20 feet in Table I.; therefore look opposite half the weight, or five tons, and the depth will be found to be 8·2 inches.

I have taken the same length and weight in each of these examples for the purpose of showing how much the depth must be increased to give stiffness.

15. When a bar or beam is employed to support a load in the middle, or at any other point of the length, a great saving of the material is made by making the bar thin and deep,[2] provided it be not made so thin as to break sideways.

The depth of a beam is sometimes limited by circumstances, and as no proportion could be given

---

[2] The term depth is always employed for the dimension in the direction of the pressure.

that would suit for every purpose, it is left entirely to the judgment of the person who may use the Table. But there is a limit to the depth, which, if it be exceeded, renders the use of cast iron for bearing purposes very objectionable and dangerous where the load is likely to acquire some degree of momentum from any cause; for if the depth be increased, it renders a beam rigid or nearly inflexible, and then a comparatively small impulsive force will break it. A very rigid beam resembles a hard body; it will bear an immense pressure, but the stroke of a small hammer will fracture it.

In order to mark the point where the depth has arrived at that proportion of the length which makes it become dangerously rigid, I have stopped the column of depths at that point, and should it be required to sustain a greater weight, the breadth must be increased instead of the depth.

16. *Example* 5. Find the depth of a rectangular bar of cast iron to support a weight of 10 tons in the middle of its length, the deflexion not to exceed one-fortieth of an inch per foot in length, and the length 20 feet; also let the depth be six times the breadth.

Under the length 20 feet in Table I. and opposite six times the weight, will be found the depth, which in this case is 15·3 inches, and the breadth will be one-sixth of this depth, or 2·6 inches.

If a fir beam be proposed to support the same weight with the same quantity of deflexion, multiply the depth 15·3 inches by 1·71, which gives 26·2 inches for the depth of the fir beam, and its breadth will be

$$\frac{26 \cdot 2}{6} = 4 \cdot 37 \text{ inches nearly.}$$

The depth of an oak beam for the same purpose may also be found by using the multiplier given for oak at the foot of the Table.

In the same manner, if the depth had been fixed to be four times the breadth, look opposite four times the weight for the depth, and make the breadth one-fourth of the depth, and so of any other proportion.

17. *Example* 6. If the breadth and length of a beam be given, and it be required to find the depth such that the beam may sustain a given weight without impairing its elastic force; then, in the second Table, the depth and deflexion may be found thus: Divide the given weight by the breadth; the quotient will be the weight a beam of one inch in breadth would sustain, which being found in the column of weights under the given length, the depth required will be opposite to it, and also the deflexion.

Let the given breadth be three inches, the weight to be supported 10 tons or 22,400 lbs., and the length 20 feet. Then

$$\frac{22400}{3} = 7466;$$

and the weight nearest to 7466 lbs. in the column

for 20 feet lengths in the second Table is 8330, and the depth 14 inches, and the deflexion would be 0·57 in.

Example 7. The second Table may be usefully applied to proportion the parts of a very simple weighing machine for weighing very heavy weights. For the flexure of a beam being directly proportional to the load upon it, while its elastic force is perfect, this flexure may be made the measure of the weight upon the beam. And a multiplying index may be easily made to increase the extent of the divisions so as to render them distinct enough for any useful purpose.

Suppose that 4 tons (8960 lbs.) is the greatest load to be weighed, and that the distance between the supports is 16 feet; and make the breadth of the bar 7 inches. Then,

$$\frac{8960}{7} = 1280,$$

and the nearest load above this under the length 16 feet in Table II. is 1328 lbs., and the corresponding depth 5 inches, which may be the depth of the bar. The flexure will be 1·02 inches, but if the beam be formed as fig. 4, Plate I., the flexure will be greater, being nearly 1·7. (The calculation may be made by art. 232.)

By making the index move over 5 inches when the deflexion is one inch, each cwt. will cause the index to move over one-tenth of an inch; but the scale should be graduated by the actual application of ton weights.

Two such beams and an index would form a simple weigh-bridge, which would be very little expense; a correct enough measure of weight for any practical use, not likely to get out of order, and would require no attention in weighing except examining the index. And this index might be enclosed, if necessary, so as to be inaccessible to the keeper of the weigh-bridge.

18. *Example* 8. To find the diameter for a mill shaft which is to be a solid cylinder of cast iron, that will bear a given pressure, the flexure in the middle not to exceed one-fortieth of an inch for each foot in length.

Let us suppose the distance of the supported points of a shaft to be 20 feet, and the pressure to be equal to 10 tons. Then multiply the pressure [3] by the constant multiplier 1·7, that is,

$$10 \times 1{\cdot}7 = 17,$$

and in this case, opposite 17 tons in the first Table, and under 20 feet, we find 11·2 inches for the diameter of the cylinder or shaft.

But a mill shaft should have less flexure than one-fortieth of an inch for each foot in length; about half that degree of flexure will be as much as should be allowed to take place. Therefore opposite double

---

[3] See art. 258, or Elementary Principles of Carpentry, Sect. II. art. 96; or edition by Mr. Barlow, 4to. 1840.

the weight, or twice 17 tons, will be found the diameter to give the shaft that degree of stiffness, that is, 13·3 inches.

If it be for a water wheel, for example, the stress should include every force acting on the shaft; that is, the weight of the wheels on the shaft, and twice the weight of water in the buckets of the water wheel; and though it will exceed the actual stress as much as the difference between the weight of the water and its force to impel the wheel, the difference is too small to render it necessary to adopt a more accurate mode of computation.

Example 9. Large shafts are often made hollow in order to acquire a greater degree of stiffness with a less weight of metal, not only to lessen the first expense, but also to lessen the pressure, and consequently friction on the gudgeons. If the thickness of the metal be made one-fifth of the external diameter, the stiffness of the hollow tube will be half that of a square beam, of which the side is equal to the exterior diameter of the tube. (See art. 259.) Therefore in Table I., opposite double the stress on the shaft, will be found the diameter in inches under the given length.

For instance, let the shaft be 25 feet long, and the stress upon it when collected in the middle 18 tons; under 26 feet in Table I. and opposite 2 × 18, or 36 tons, will be found 15·3 inches, the diameter of the shaft, provided it may bend 0·65 in., or a little more than half an inch at every revolution. If it should bend only half this, then look opposite twice 36 tons; the nearest in the Table is 75 tons, and the diameter is 18½ inches. The thickness of metal will be one-fifth, or nearly 3¾ inches.

19. *Example* 10. When the diameter of a solid cylinder is given, and the length, to find the greatest load it will sustain without injury to its elasticity, and the deflexion that weight will cause.

Suppose the diameter to be 11 inches, and the length 20 feet, then in the second Table, opposite the depth 11 inches, and under the length 20 feet, will be found 5142 lbs. Let this be multiplied by the diameter 11 inches, and divided by the constant number 1·7; the result will be the weight required in pounds.

In this case it is 33,271 lbs., for

$$5142 \times 11 \div 1{\cdot}7 = 33{,}271.$$

The deflexion opposite 11 inches and under 20 feet is ·73 in.

Any different degree of deflexion may be allowed for in the same manner as shown in the third and fourth examples.

APPLICATION TO CASES WHERE THE LOAD IS TO BE UNIFORMLY DISTRIBUTED OVER THE LENGTH OF THE BEAM.

20. Whether a load be uniformly distributed over the length from A to B, fig. 2, Plate I., or it be collected at several equidistant points, as at 1, 2, 3, 4,

5, 6, and 7, in the same figure, the same rule may be used, as it causes no difference that need be regarded in practice. But the effect of this load in producing flexure differs from its effect in producing permanent alteration.

It is proved by writers on the resistance of solids, that the whole of a load upon a beam, when it is uniformly distributed over it, will produce the same degree of deflexion as five-eighths of the load applied in the middle,[4] (see experiment, art. 54, 61, and 62). Consequently, take five-eighths of the whole load upon the beam, and with this reduced weight proceed as in the foregoing examples.

21. *Example* 11. Let it be required to find the dimensions of a cast iron bar to support 10 tons uniformly distributed over its length, the depth of the bar to be four times its breadth, and the deflexion to be not more than one-eightieth part of an inch for each foot in length, or one-fourth of an inch, the length being 20 feet.

Here the five-eighths of 10 tons is 6 tons and a quarter, and as the depth is to be four times the breadth, multiplying six and a quarter by four gives 25 tons; but the deflexion is to be only half that given in the Table; therefore the 25 must be doubled, which gives 50 for the number of tons

---

[4] Dr. Young's Lectures on Nat. Phil. vol. ii. art. 325, 329. Mr. Barlow's Treatise on the Strength of Timber, Cast Iron, &c., art. 55. 1837.

opposite which the depth is to be found. The depth opposite 50 tons, and under 20 feet, is 14·6 inches, and the breadth is

$$\frac{14 \cdot 6}{4} \text{ or } 3 \cdot 65 \text{ inches;}$$

that is, a bar 14·6 inches deep, and 3·65 inches in breadth, will bear a load of ten tons uniformly distributed over it when the length of bearing is 20 feet, and the deflexion in the middle a quarter of an inch.

Example 12. Let it be proposed to find the proper dimensions for an open girder of cast iron, for supporting the floor of a room, the girder being formed as described in fig. 11, Plate II. (See art. 41.)

Suppose the distance between the walls to be 25 feet, and the distance between girder and girder to be 10 feet, then there will be

$$10 \times 25 = 250$$

superficial feet of floor supported by each girder; and the load on each foot being 160 lbs., (see Alphabetical Table, art. Floor,)

$$160 \times 250 = 40{,}000 \text{ lbs.}$$

is the whole load distributed over the girder. But five-eighths of 40,000 is 25,000 lbs., and multiplying[5] 25,000 by 6·3 we have 157,000 lbs.; the nearest

[5] It is shown in a note to art. 200, that where the breadth and depth of the section of the beam at A B, or C D, fig. 11, is one-fifth of the entire depth of the beam in the middle, the strength is to that of a square beam as 1 : 6·3, and the stiffness is in the same proportion.

number in Table I. is 156,800 lbs., and the mean between the depths for 24 and 26 feet is 17·8 inches, which is the depth for the girder: the breadth should be one-fifth of the depth, or

$$\frac{17 \cdot 8}{5} = 3 \cdot 56 \text{ inches,}$$

and the section at A B, and C D, square.

If the girder were actually loaded to the extent we have calculated upon, the depression in the middle would be about one-third more than is stated at the foot of the Table, in consequence of the girder being diminished towards the ends; but the greatest variable load in practice is seldom more than half that we have assumed, and it is the flexure from the variable load which is most injurious to ceilings, &c.

Again, let the length of bearing be 20 feet, and the distance of the girders 8 feet, and the weight 160 lbs. upon a superficial foot of the floor, then

$$20 \times 8 \times 160 = 25{,}600 \text{ lbs.}$$

the whole load distributed over the girder. And five-eighths of this load multiplied by 6·3 is

$$\frac{5 \times 6 \cdot 3 \times 25{,}600}{8} = 100{,}800 \text{ lbs.}$$

The nearest number in Table I. is 103,040, and the depth corresponding to a 20 feet bearing is 14·3 inches, the depth of the girder required; and

$$\frac{14 \cdot 3}{5} = 2 \cdot 86 \text{ inches,}$$

the breadth.

The examples here given of girders show the dimensions of some that were executed several years ago.

Example 13. The same calculations apply to the form of girder shown in fig 24, Plate III. When the extreme breadth at the upper or lower side is one fifth of the depth, divide this breadth into ten equal parts, and make the thickness in the middle of the depth four of these parts; the depth of the projections should be three-fourths of the breadth.[6] With these proportions, the depth at the middle of a girder for a 25 feet bearing should be 17 8 inches, and the extreme breadth 3 56 inches, as in the preceding example.

And for a 20 feet bearing $14\frac{3}{10}$ inches deep, and 2·86 inches in breadth. I have seen some of less dimensions employed in several instances, but it is to be hoped such examples are not very common. A review of my mode of calculation will show that no more excess of strength is allowed than ought to be in such a material.

When there is not any length and weight in the Table exactly the same as those which are given, take the nearest; the dimensions thus obtained will always be sufficiently near for practice.

22. In applying the second Table, the effect of a load uniformly distributed over the length is to be considered equal to that of half the load collected at

---

[6] See note to art. 186, for the reason of this rule.

the middle point, (art. 139.) Therefore considering this half load the weight to be supported, proceed as in the other examples of the use of the second Table.

EXAMPLES OF THE USE OF THE THIRD TABLE.

23. *Example* 14. Let it be required to support the floor of a warehouse by iron pillars, where the greatest load on any pillar will be 70 tons, the height of the pillars being 14 feet.

Seventy tons is equal to 1400 cwt.; and in the column having 14 feet at the head, in the third Table, 1561 cwt. is the nearest weight; and the diameter opposite this weight in the side column is 9 inches, the diameter required.

If it be wished to approach nearer to the proportion, take the mean between the weight above and that below 1400; that is, the mean between 1561 and 1185, which is 1373, or nearly 1400; hence it appears that a little more than $8\frac{1}{2}$ inches would be a sufficient diameter, but it is seldom necessary to calculate so near.

Example 15. If it be desired to fix on the diameter for story posts of cast iron to support the front of a house; such a one for example as is commonly erected in London where the ground story is to be occupied with shops;—in such a case, each foot in length of frontage may be estimated at 25 cwt. for each floor, and 12 cwt. for the roof: hence in a house with three stories over the shops, the extreme load will be

$$\overline{3 \times 25} + 12 = 87 \text{ cwt.}$$

on each foot of frontage. Now if the posts be 7 feet apart, and 12 feet high, we have $7 \times 87 = 609$ cwt. the load upon one post; and hence we find by the Table, that a pillar $6\frac{1}{2}$ inches in diameter would be sufficient; the load 525 cwt., which corresponds to a diameter of 6 inches, being too small

If there be only two stories above the pillars, and the height of a pillar be 10 feet, the distance from pillar to pillar 7 feet; then,

$$\overline{(2 \times 25) + 12} \times 7 = 434 \text{ cwt.}$$

the whole load for one pillar: and it appears by the Table, that a pillar 5 inches in diameter would sustain 452 cwt.; consequently 5 inches will be a proper diameter for the pillars.

When pillars are placed at irregular distances, that which carries the greatest load should be calculated for, and if it happen that such a pillar stands 10 feet from the next support on one side, and 6 feet from the next support on the other side, add these distances together, and take the mean for the distance apart; thus,

$$\frac{10 + 6}{2} = \frac{16}{2} = 8,$$

the mean distance of the supports.

The strain upon a pillar cannot be exactly in the direction of the axis when the pillars are placed at unequal distances to support an uniform load; and since this unequal distribution of supports is extremely common in story posts, the propriety of adopting the mode of calculation I have followed is evident.

The diameter of a story post is sometimes made so small in respect to its height and the load upon it, that a very slight lateral stroke would break it: while we hope that no serious accident may occur through such hardihood, we cannot but dread the consequences of trusting to these inadequate supports.

SECTION III.

OF THE FORMS OF GREATEST STRENGTH FOR BEAMS.

24. In the Introduction, I have stated that one of the most valuable properties of cast iron consists in our being able to mould it into the strongest form for our intended purpose; and in order to apply this property with the most advantage, it will be useful to consider the means of applying our theoretical knowledge on this subject to practice.

There are two means of increasing the strength of a beam; the one consists in disposing the parts of the cross section in the most advantageous form; the other, in diminishing the beam towards the parts that are least strained, so that the strain may be equal in every part of the length.

OF FORMS OF EQUAL STRENGTH FOR BEAMS TO RESIST CROSS STRAINS.

25. Before I point out the forms of equal strength corresponding to different modes of applying the load or straining force, let us consider the condi-

tions that are essential in a practical point of view. In the first place, supported parts must have sufficient magnitude to insure stability; for it is much more important that every connexion or joining should be firm, and that the bearing parts should be secure against crushing or indentation, than it is that a small portion of material should be saved. When mathematicians investigate a form of equal strength, the manner of connecting it or supporting it is not considered. Girard has shown that whatever line generates a solid of equal resistance, the solid always terminates in a simple point, or in an arris which is either perpendicular or parallel to the direction of the straining force.[1] Therefore the forms given by this mode of investigation do not answer in practice unless they be properly modified.

26. It may be easily proved, that in a rectangular section, when a weight is supported by a beam, the area of the section at the point of greatest strain should be to the area at the place of least strain, as six times the length is to the depth at the point of greatest strain;[2] and this is the least pro-

---

[1] Traité Analytique de la Résistance des Solides, art. 129.

[2] For it is shown (art. 110) that

$$\frac{f\,b\,d^2}{6\,l} = \mathrm{W},$$

but the force to resist detrusion being as the area simply; therefore we must have $f\,b'\,d' = \mathrm{W}$ at the weakest point. Consequently

portion that ought to be given. Now when the length and depth are equal, the area at the point of least strain should be one-sixth of the area at the point of greatest strain, instead of being a simple point or an arris.

27. If a beam be supported at the ends, and the load applied at some one point between the supports, and always acting in the same direction, the best plan appears to be to keep the extended side perfectly straight, and to make the breadth the same throughout the length; then the mathematical form of the compressed side is that formed by drawing two semiparabolas A C D and B C D, fig. 3, C being the point where the force acts.[3] Now the curve terminating at A, it is necessary in applying it to use, to add some such parts as are indicated by the dotted lines at the extremities. The same form is proper for a beam supported in the middle, as the beam of a balance.

28. Irregular additions of this kind, however, render it difficult to estimate the effect of the straining force; therefore, some simple straight-lined figure to include the parabolic form is to be pre-

---

$$\frac{b\,d^2}{6\,l} = b'\,d';$$

or

$$6\,l : d :: b\,d : b'\,d';$$

where l is the length, $b\,d$ the area at the point of greatest strain, and $b'\,d'$ the area at the point of least strain.

[3] Greg. Mechanics, i. art. 180. It was first shown by Galileo.

ferred: this may be easily effected as proposed by Dr. Young,[4] by making the lines bounding the compressed side tangents to the parabolas, as in fig. 4. If A E be equal to half C D, then E C is a tangent to the point C of an inscribed parabola A C, having its vertex at A.

By forming a beam in this manner, one-fourth of the material is saved; but the flexure will be somewhat more than one-third greater, therefore there is a loss of stiffness in using this form.

29. If the beam be strained sometimes from one side and sometimes from the other, both sides should be of the same figure, as in fig. 5. In the beam of a double acting steam engine, the strain is of this kind. A E and B F should be equal, and each equal to half C D as before.

30. It is sometimes desirable to preserve the same depth throughout; and in this case, the section through the length of the beam made perpendicular to the direction of the straining force should be a trapezium, described in the manner shown in the 6th figure,[5] the force acting perpendicularly at C, the points of support being at A and B. A figure of this kind would obviously be without stability, but modified as shown by fig. 7, the end being formed as shown at B′, any degree of stability may be given, and with a less quantity of material than

[4] Nat. Philos. vol. i. p. 767.

[5] Gregory's Mechanics, i. art. 179.

when the depth is diminished, as in the parabolic form. Also, the deflexion is less, which gives this form a considerable advantage for bearing purposes. In a beam supported in the middle, the same form may be used when the weights act at the ends, as in a balance.

31. When a beam or bar is regularly diminished towards the points that are least strained, so that all the sections are similar figures, whether it be supported at the ends and loaded in the middle, or supported in the middle and loaded at the ends, the outline should be a cubic parabola;[6] and if the section of the beam be a circle at the point of greatest strain, the form of the beam should be that generated by the revolution of the cubic parabola round its axis, the vertex being at the point of least strain.

But in practice, a frustum of a cone or a pyramid will generally answer better, the diameter of the point of greatest strain being to that at the point of smallest strain as 3 : 2.[7]

The same figure is proper for a beam fixed at one end, and the force acting at the other; consequently, it is a proper figure for a mast to carry a single sail.

---

[6] Gregory's Mechanics, i. art. 181, or Emerson's Mechanics, prop. lxxiii. cor. 1.

[7] Such a cone or pyramid will include the figure of equal strength, the subtangent of the curve being three times its abscissa.

32. If a weight be uniformly distributed over the length of a beam supported at both ends, and the breadth be the same throughout, the line bounding the compressed side should be a semi-ellipse when the lower side is straight,[8] as shown in fig. 8.

Instead of an ellipse, I usually make the compressed side a portion of a circle, of which the radius is equal to the square of half the length divided by the depth of the beam. The dotted line in fig. 8 shows this form.

The same form of equal strength should be employed when the beam is intended to resist the pressure of a load rolling over it; hence the beams of a bridge, the rails of a waggon-way, and the like, should be of this figure.

33. If a beam has to bear a weight uniformly distributed over its length, and its depth be every where the same, the beam being supported at both ends, then the outline of the breadth should be two parabolas A C B, A D B, set base to base, their vertices C and D being in the middle of the length, as shown in the annexed perspective sketch.[9] In prac-

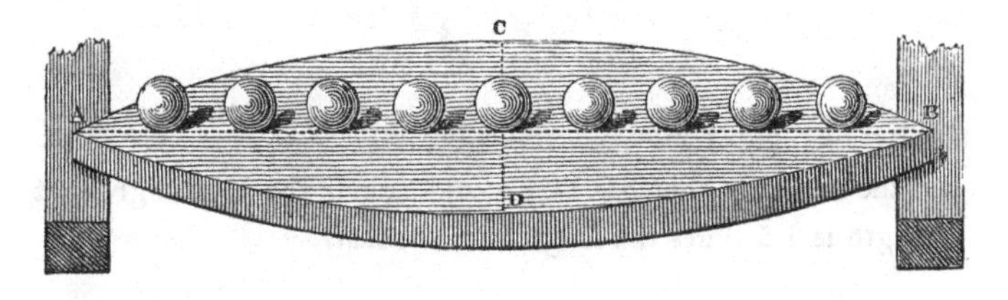

[8] Gregory's Mechanics, i. art. 182, or Emerson's Mechanics, prop. lxxiii. cor. 3.

[9] Young's Nat. Phil. i. p. 767.

tical cases, the arcs A C B, A D B, may be portions of circles.

When the ends are modified as in fig. 7, Plate I., this will be the most advantageous form for a beam for supporting a load uniformly distributed over its length, as lintels, bressummers, joists, and the like.

34. When a beam is fixed at one end only, and has to support a weight uniformly distributed over its length; if the breadth of the beam be every where the same, the form of equal strength is a triangle A C B,[10] fig. 21, Plate III.

35. If a beam be fixed at one end only, and the weight be uniformly diffused over the length, the section being every where circular, then the form of equal strength would be that generated by the revolution of a semi-cubic parabola round its axis.[11]

It will be sufficient in practice to employ the frustum of a cone of which the diameter at the unsupported end is one-third of the diameter at the fixed end.[12]

---

[10] Emerson's Mechanics, prop. lxxiii. cor. 2. [11] Ibid.

[12] For the equation of the curve is,

$$a\,x = y^{\frac{3}{2}};$$

hence,

$$\tfrac{3}{2}\,x = \text{the subtangent} = 1{\cdot}5\,x;$$

and the length of the cone that would include the form of greatest strength is 1·5 times the length of the beam.

SECTION IV.

OF THE STRONGEST FORM OF SECTION.

36. When a rectangular beam is supported at the ends, and loaded in any manner between the supports, it may be observed that the side against which the force acts is always compressed, and that the opposite side is always extended; while at the middle of the depth there is a part which is neither extended nor compressed; or, in other words, it is not strained at all.

Any one who chooses to make experiments may satisfy himself that this is a correct statement of the fact, in any material whatever; whether it be hard and brittle as cast iron, zinc, or glass; or tough and ductile as wrought iron and soft steel; or flexible as wood and caoutchouc; or soft and ductile as lead and tin. In very flexible bodies it may be observed by drawing fine parallel lines across the side of the bar before the force is applied; when the piece is strained the lines become inclined, retaining their original distance apart only at the neutral axis. In almost all substances, the fracture shows distinctly that a part has been extended, and the rest compressed; and in some substances, as wood, lead, tin, wrought iron, &c., the place of the axis of motion may be observed in the fracture. It was first noticed in experiment, and applied to correct Galileo's theory by Marriotte.[1] Coulomb[2] and Dr. Young have made it the basis of their investigations, the latter showing the important fact that an oblique force changes the position of this axis;[3] as has been investigated more in detail in this Essay. Lately the place of the neutral axis in horizontal beams has been more closely examined by Barlow in a numerous course of experiments;[4] and Duleau has ingeniously shown its place by experiment on wrought iron.[5] The same thing is exhibited in a refined and beautiful contrivance of Dr. Brewster's, which he calls a teinometer, and employs to measure the effect of force on elastic bodies.[6]

The strains decrease from each side towards the middle, and in the middle they are insensible. I

---

[1] Treatise on the Motion of Water, &c., translated by Desaguliers, p. 243, 8vo. London, 1718.

[2] Mémoires de l'Académie des Sciences. Paris, 1773.

[3] Lectures on Natural Philosophy, vol. ii. p. 47, 4to. London, 1805.

[4] Essay on the Strength of Timber, &c., p. 88, &c., 8vo. London, 1817. Since reprinted, with additions, in 1838.

[5] Essai Théorique et Expérimental sur la Résistance du Fer Forgé, p. 26, 4to. Paris, 1820.

[6] Additions to Ferguson's Lectures, vol. ii. p. 232, 8vo. Edinburgh, 1823.

will call the part at the middle of the depth the neutral axis, or *axis of motion*. See Sect. VII. art. 107.

37. In the case of equilibrium, between the straining force and the resistance of a beam, it is a necessary condition that the resistance on one side of the axis of motion should be exactly equal to the resistance on the other side; or, that the force of compression should be equal to the force of extension. Now, in practice, a body should never be strained beyond its power of restoring itself; and as it is known from experience, that while their elastic force remains perfect, bodies resist the same degree of extension or of compression with equal forces, it is obvious that, in the section of a beam of the greatest strength, the form on each side of the axis of motion should be the same; because whatever is the strongest form for one side of the axis must be equally so for the other. Hence, the axis of motion in beams of the greatest strength will always be at the middle of the depth.[7]

38. And, as it is shown by writers on the resistance of solids, that the power of any part in the same section is directly as the square of its distance from the axis of motion, (art. 108,) when the strain

---

[7] These remarks apply only to bodies subjected to very moderate strains, particularly in cast iron; since that metal requires, on the average, nearly seven times as much force to crush it as to tear it asunder, and the breaking strength of beams depends upon these forces.—EDITOR.

upon it is the same, it is obviously an advantage to dispose the parts of the section at the greatest possible distance from the axis of motion, provided that the middle parts be kept sufficiently strong to prevent the straining force from crushing the extreme parts together, and that the breadth be made sufficient to give stability.

39. It must also be observed that when the parts are not of equal thickness, the metal cools unequally, and therefore is partially strained by irregular contraction; it is sometimes even fractured by such irregular cooling: for this reason, the parts of a beam should be nearly of the same size. A good founder may generally reduce the danger of irregular cooling, but it is always best to avoid it altogether.

40. The form of section which I usually adopt in order to fulfil these conditions is represented in fig. 9.[8] A M is the axis of motion; the parts on each side of the axis of motion are the same; the metal is nearly of equal thickness, and the parts necessary to give strength and stability are disposed at the greatest distance from the axis of motion.

A section of this form is adapted for many purposes; such, for example, as the beam of a steam engine, as in fig. 26; or for supporting arches, as in fig. 10, for girders, bearing beams, and the like.

[8] This is not the form of greatest strength to resist fracture; and the beam proposed in the next article (fig. 11, Plate II.) breaks irregularly, and is remarkably weak. See Additions.—EDITOR.

41. When it is necessary to leave some part of the middle of the beam quite open, or when the depth is considerable, I have recourse to another method, which has, in such cases, a decided advantage in point of economy. It consists in making the compressed side of the beam, or that against which the force acts, a series of arches, and the other side a straight tie. (See fig. 11, Plate II.) If the tie be not straight, there is a great loss of strength, and a greater loss of stiffness.

In this figure, the thickness is supposed to be every where the same, and the narrowest part of the curved side of the same width as the straight side; or, so that the area of the section at A B may be the same as the area of the section at C D.

The sketch in the figure is for the case in which the load is uniformly distributed over the length, and then the upper side should be the proper curve of equilibrium for an uniform load. This curve is a common parabola, but a circular arc will always be sufficiently near when the rise is so small. The upper part of the beam forms an arch, of which the continued tie forms the abutments, and the smaller arches are merely to connect the two parts and give stability to the whole.

The connexion thus formed is necessary for supporting the tie; and in consequence of this connexion the effect of the straining force will be similar to that on a solid beam. Several girders and beams for floors have been formed on this principle; and a simple method of proportioning them will be found in the second Section, p. 41.

All the parts should be kept as nearly as convenient of the same bulk, to prevent irregular contraction.

42. If the load be distributed in any other manner, the curve should be the proper curve of equilibrium for that load.[9]

For if it be not the proper curve, partial strains will be produced in the beam, which will impair its strength. The curve of equilibrium should pass every where at the middle of the depth of the curved part of the beam, and should meet the axis of the straight tie in the centres of the supports upon which the beam rests. Thus A C being the curve of equi-

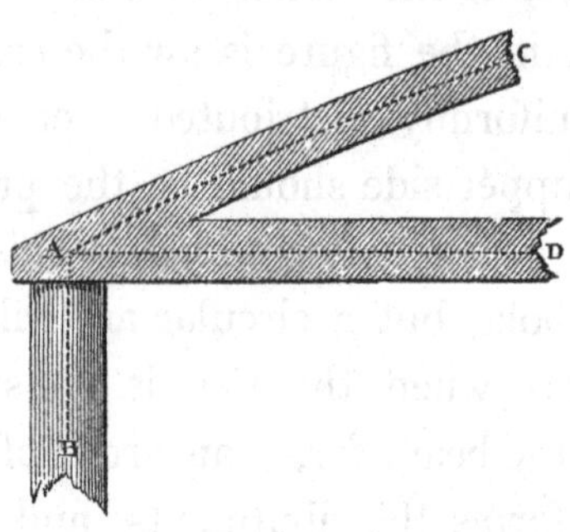

librium, A D the axis or centre line of the tie; A B should be the centre of the support on which the beam rests.

43. If the load be applied at one point, the upper side should be formed of two straight lines, meeting

[9] The method of finding the curve of equilibrium is shown in my "Elementary Principles of Carpentry," Sect. I. art. 47-61.

in the point where the load is to rest, as at A in fig. 12.

The openings should be disposed as may best answer the purpose for which the beam is intended, but they may generally be from 2 to 3 feet each. When such beams, as fig. 11, are used as girders, the openings receive the binding joists instead of mortises.

44. When a beam is to bear a load at one end, the other being fixed; or when a beam is loaded at both ends and the support is in the middle; then the tie should be the upper part of the beam: it should obviously be straight for the reasons already assigned; and the other parts should be straight also, except the small degree of curvature which would cause the weight of the part to be balanced by the forces concerned. Indeed, the arrangement for this strain should be the same as fig. 12 inverted, the support being at A, and the load at B and D.

45. But when the load is uniformly distributed over the length, the lower side A C, in the annexed figure, should be curved; the proper curve for an

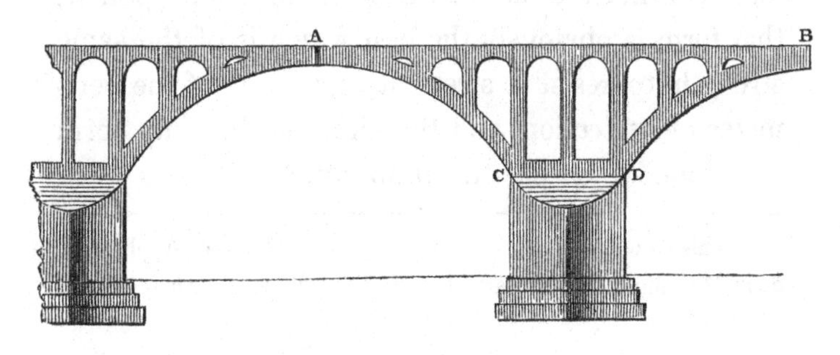

uniform load being a common parabola with its vertex at A. By a combination of such beams, a bridge might be formed which would have no lateral pressure on its piers or abutments. C D being one of the piers, the distance between the points C and D may very easily be so arranged, that a given force at A or B would not disturb the equilibrium of the frame.

A bridge of this kind would not be affected by contraction and expansion; because no connexion would be necessary at the junction of the beams at A, but such as would allow of the motion of contraction or expansion without injury.

In a design for a large bridge on this principle, which I made some years ago, it was contrived to put together in parts, without the assistance of centering; the open work of the spandrils being composed of vertical and diagonal supports and braces.

OF THE STRONGEST FORM OF SECTION FOR REVOLVING SHAFTS.

46. When a beam revolves, while the straining force continues to act in the same direction upon it, that form is obviously the best which is of the same strength to resist a stress at any point of the perimeter of its section, and the circle is the only form of section which has this property.[10]

---

[10] This conclusion has been objected to by Navier (Application de la Mécanique, note to art. 494) in the following terms :—

If a shaft be of any other form than cylindrical, the flexure will be different in different parts of the revolution, and therefore the motion will be unsteady, and particularly in new work. In a square shaft, (and such shafts are chiefly employed,) the resistance to pressure at one point is to the resistance to the same pressure at another point in the perimeter as ten is to seven nearly (art. 112). In feathered shafts, that is, shafts of which the section is similar to fig. 13, the resistance is more regular, but not perfectly so.[11]

"The most convenient figure for axes of rotation is made a subject of inquiry in the Practical Essays on Mill-work by Buchanan, with notes by T. Tredgold, 1823, vol. i. p. 262, and in the Practical Essay on the Strength of Cast Iron. Mr. Tredgold appears to think that the circle is the only figure which gives to the axes the property of offering in every direction the same resistance to flexure. The error of this engineer proceeds from his considering the resistance to flexure as being measured by the expression which measures the resistance to rupture. We have already remarked that a square section gave the same resistance to flexure in the direction of the sides and of the diagonals. But moreover, this section gives an equal resistance in every direction; and the same is the case with regard to a great number of figures, which may be formed by combining in a symmetrical manner the circle and the square. It thence results that if the axes strengthened by salient sides, which the English call feathered shafts, do not answer as well as square axes, or full cylindrical ones, this arises probably from their not being as well disposed to resist torsion, and not from the inequalities of flexure of these axes."—EDITOR.

[11] In heavy astronomical instruments, and in all machines where steady and accurate movements are necessary, every atten-

For the same reasons, the sections of the masts of vessels should be circular.

47. As the circle is the best form for the section of a shaft, a hollow cylinder will be the strongest and stiffest form for a shaft; and the same form is also best calculated for resisting a twisting strain to which all shafts are more or less exposed.

The idea of making hollow tubes for resisting forces that often change their direction, has been undoubtedly borrowed from nature; but in art we cannot pursue the principle to so much advantage, because it is difficult to make a perfect casting of a thin tube; and in shafts, &c. of small diameter, it is much greater economy to make them solid.

It is usual to make hollow tubes of uniform diameter with gudgeons cast separate, to fix at the ends. The manner of calculating the stiffness of hollow tubes for shafts is shown in art. 259 and 260, and an easy popular mode at art. 18. When they are applied to other purposes, consult art. 178, and those following it in the same proposition.

---

tion should be paid to the effect of flexure. Irregularity may be diminished by excess of strength, but it cannot be wholly removed. The reader who wishes to pursue this subject, as far as regards astronomical instruments, may consult the Philosophical Magazine, vol. lx. p. 338, and vol. lxi. p. 10.

SECTION V.

AN ACCOUNT OF SOME EXPERIMENTS ON THE RESISTANCE OF CAST IRON.

48. There have been very few experiments made on the resistance of cast iron, in which the degree of flexure produced by a given weight has been measured; but the few that have come to my knowledge, and that are sufficiently described to admit of comparison, I purpose to compare with the rules I made use of in calculating the Tables in this work; and to add several new experiments.

MR. BANKS'S EXPERIMENTS.[1]

49. Mr. Banks made some experiments on cast iron, and noticed the deflexion, but only at the time of fracture. These experiments were made at a foundry at Wakefield. The iron was cast from the air-furnace; the bars one inch square, and the

[1] From a treatise "On the Power of Machines," by John Banks. Kendall, 1803, p. 96.

props exactly a yard distant. One yard in length weighed exactly 9 lbs., excepting one, which was about half an ounce less, and another a very little more. They all bent about an inch before they broke.

1st bar broke with	. . . .	963 lbs.	Mean $971\frac{2}{3}$ lbs.
2d bar broke with	. . . .	958 ,,	
3d bar broke with	. . . .	994 ,,	
4th bar, made from the cupola, broke with		864 ,,	

50. Now the rule according to which the first Table was calculated is expressed by the equation

$$\cdot 001\ W L^2 = B D^3,$$

in which the weight in pounds is denoted by W, the length in feet by L, the breadth in inches by B, the depth in inches by D, and the number ·001 is a constant multiplier, which I shall sometimes denote by a.

The rule determines the dimensions for a deflexion of as many fortieths of an inch as there are feet in length, or $\frac{L}{40}$; and if d be the deflexion in inches determined by experiment, we have

$$d : W :: \frac{L}{40} : \frac{W L}{40\, d},$$

which being substituted for the weight in the equation above it, becomes

$$\frac{\cdot 001\ W L^3}{40\, d} = B D^3,$$

$$\text{or, } \cdot 001 = a = \frac{40\ B D^3 d}{W L^3}.$$

The equation, in this form, may be called a formula of comparison, as when the value of a determined by it is the same I have used, or nearly the same, it will be evident that the Table is calculated from proper data.

51. Taking the mean of the first three of Mr. Banks's experiments, we have

$$\frac{40\,B\,D^3\,d}{W\,L^3} = \frac{40}{971 \times 27} = \cdot 00152 = a.$$

And in the bar from the cupola, or fourth experiment,

$$\frac{40\,B\,D^3\,d}{W\,L^3} = \frac{40}{864 \times 27} = \cdot 0017 = a.$$

The experiments of Mr. Banks indicate therefore that he had employed iron of a more flexible quality, but they are not sufficiently accurate for establishing the elements of a practical rule, because the deflexion was not correctly observed, nor observed at a proper stage of the experiment. For when a bar is strained nearly to the point of fracture, the deflexion becomes extremely irregular, and increases more rapidly than in the simple proportion of the weight, (see art. 56, 63, 65, and 67,) and consequently must give a much higher value to a than the true one, as we find to be the case with these experiments.

M. RONDELET'S EXPERIMENTS.[2]

52. M. Rondelet has described some experi-

[2] Extracted from his Traite Théorique et Pratique de l'Art de Bâtir, 6 tomes, 4to. Paris, 1814, tome iv. p. 514.

ments on different kinds of cast iron in his work on building, which were made upon specimens of 1·066 inches square, supported at the ends, and loaded in the middle of the length.

M. Rondelet's First Experiments. Distance between the Supports 3·83 *feet.*

Weight in lbs.	134	201	268	335	Remarks, &c.
Kind of iron.	Defl. inch.	Defl. inch.	Defl. inch.	Defl. inch.	
1. Gray cast iron	·089	·2	·357	·49	Broke with 482 lbs.
2. Do. do.	·156	·313	·38	·49	Broke with 482 lbs.
				2) ·98	·49 mean of deflexions, with 335 lbs.
3. Soft cast iron	·134	·313	·466	·62	Broke with 700 lbs.
4. Do. do.	·0223	·067	·134	·2	Broke with 1140 lbs.
5. Do. do.	·089	·156	·245	·38	Broke with 375 lbs.
6. Do. do.	·089	·178	·29	·445	Broke with 605 lbs.
				4) 1·645	·411 mean of deflexions, with 335 lbs.

M. Rondelet's Second Experiments. Distance between the Supports 1·915 *feet.*

Weight in lbs.	322	483	644	805	Remarks, &c.
Kind of iron.	Defl. inch.	Defl. inch.	Defl. inch.	Defl. inch.	
1. Gray cast iron	·067	·089			Broke with 580 lbs.
2. Do. do.	·0445	·089	·112	·134	Broke with 1063 lbs. Mean of deflexions, with 483 lbs. is ·089 inch.
3. Soft cast iron	·0445	·089	·134	·153	Broke with 1770 lbs.
4. Do. do.	·0445	·067	·134		Broke with 1360 lbs.
		2) ·156			Mean of deflexions, with 483 lbs. is ·078 inch.
		·078			

In order to compare these results with the formula used in calculating the Tables, I have taken the mean deflexions corresponding to the load of 335 lbs. in the long pieces, and to 483 lbs. in the short ones; and in the gray cast iron,

For the long lengths . . . . . $a = \cdot00134$
For the short lengths . . . . . $a = \cdot00135$

In the soft cast iron,

For the long lengths . . . . . $a = \cdot00112$
For the short lengths . . . . . $a = \cdot00118$

These values of a were calculated by the formula of comparison given in art. 50, and the latter ones nearly agree with that employed to calculate the Table.

MR. EBBELS'S EXPERIMENT.

53. According to a trial communicated to me by Mr. R. Ebbels, a bar of cast iron, 1 inch square, and supported at the ends, the distance of the supports being 3 feet, the deflexion in the middle was $\frac{3}{16}$ths of an inch, with a weight of 308 lbs. suspended from the middle. The iron was of a hard kind, not yielding very easily to the file; it was cast at a Welsh foundry.

In this trial we have

$$\frac{40\,\mathrm{B}\,\mathrm{D}^3\,d}{\mathrm{L}^3\,\mathrm{W}} = \frac{40 \times 3}{27 \times 308 \times 16} = \cdot000902 = a.$$

Consequently, iron of this kind is about $\frac{1}{10}$th stronger than that which the Table is calculated from, or rather it would bend $\frac{1}{10}$th part less under the same strain.

Experiment 1.

54. A joist of cast iron of the form described in fig. 9, Plate I., was submitted to the following trials. It was supported at the ends only; the distance between the supports 19 feet, and placed on its edge. The deflexion from its own weight was $\frac{3}{40}$ths of an inch.

When it was laid flatwise, the deflexion from its own weight was 3·5 inches, the distance of the supports remaining 19 feet.

The whole depth *a d*, fig. 9, was 9 inches, the breadth, *a b*, was 2 inches; the depth of the middle part, *e f*, was $7\frac{1}{2}$ inches; and the breadth of the middle part $\frac{3}{4}$ths of an inch.

55. It may be easily shown that to derive the value of *a*, from the experiment on the edge, we may use an equation of this form, (see art. 192 and 215,)

$$a = \frac{40\,B\,D^3\,d\,(1 - p^3\,q)}{\frac{5}{8}\,W\,L^3} = \frac{64\,B\,D^3\,d\,(1 - p^3\,q)}{W\,L^3};$$

in which D is the whole depth, and $p\,D$ the depth of the middle part, and B the whole breadth, and $q\,B$ the breadth after deducting that of the middle part.

In our experiment $D = 9$ inches, and $p\,D = 7{\cdot}5$, or $p = {\cdot}833$. Also, $B = 2$ inches, and deducting $\frac{3}{4}$ths, the breadth of the middle, we have $q\,B = 1{\cdot}25$, or $q = \cdot 625$. And the weight of the part of the joist between the supports being 540 lbs., we find $a = \cdot 00124$.

The equation for finding the value of a, in the experiment with the joist flatwise, is

$$\frac{64 \, B \, D^3 \, d \, (1 + p^3 q)}{W \, L^3} = a = \cdot 00092.$$

Where

$$D = 2 \text{ inches, } B = 9 - 7\cdot 5 = 1\cdot 5, \, p = \frac{\cdot 75}{2}, \text{ and } q = \frac{7\cdot 5}{1\cdot 5}.$$

I consider the value of a derived from the experiment with the joist flatwise as nearest the truth, because the deflexion was so considerable, that a small error in measuring it would not sensibly affect the result, while there must be some uncertainty in ascertaining so small a deflexion as $\frac{3}{40}$ths of an inch in 19 feet; and a very small error in this measure would cause the difference between the results. I have, however, given it, as I determined it at the time, and the manner of calculation may be useful in other cases. If the mean be taken between the results, it is

$$\frac{\cdot 00124 + \cdot 00092}{2} = \cdot 00108.$$

In the experiment flatwise, we obtain a constant multiplier extremely near to that determined from a bar of the same iron an inch square and 34 inches long (art. 57), and it differs only about $\frac{1}{12}$th part from the one employed for calculating the Table, page 12, art. 5.

Experiment 2.

56. I now purpose describing the direct experiments I have made for obtaining the constant multipliers used in this work; I call experiments direct when known weights are applied as the straining force, without the intervention of mechanical powers, without loss of effect from friction, or a risk of error in estimating the quantity of force, when the yielding of the supports cannot affect the measure of the deflexion, and when the deflexion can be accurately measured.

Thé iron I used was soft gray cast iron; it yielded easily to the file, and extended a little under the hammer, before it became brittle and short.[3]

The first experiment was made with a bar of an inch square, cast by Messrs. Dowson, London, with the supports 34 inches apart; the weights were placed in a scale suspended from the middle of the length; the load was increased by 10 lbs. at a time, and the deflexion measured each time, the quantity of deflexion being multiplied by means of a lever index. The whole time of making the experiment was nearly four hours; the thermometer varying from 65 to 66 degrees. Only half the number of observations is inserted here.

---

[3] A considerable degree of malleability is a good quality in cast iron for bearing purposes, because it lessens the risk of sudden failure. The iron was a mixture of Butterly iron, two parts, with one part of old iron.

Weight in lbs.	Defl. in inch.	Remarks.	Weight in lbs.	Defl. in inch.	Remarks.
20[4]	·02		240	·13	
40	·03		260	·14	
60	·04		280	·15	
80	·05		300	·16	unloaded, and it returned to its natural state.
100	·06		320	·17	
120	·07		340	·18	
140	·08		360	·19	
160	·09		380	·2	
180	·10	unloaded, and it returned to its natural state.	400	·21	
200	·11		410	·22	deflexion became irregular; and when the load was removed, it had taken a permanent set, with a curvature of ·015 inch.
220	·12				

From this experiment we find that the deflexion of cast iron is exactly proportional to the load, till the strain arrives at a certain magnitude, and it then becomes irregular; and at or near the same strain a permanent alteration takes place in the structure of the iron, and a part of its elastic force is lost. The same thing occurs in experiments on other metals: I have tried wrought iron, tin, zinc, lead, and alloys of tin and lead, with a view to measure their elastic forces, and the strains that produce permanent alteration.

57. According to this experiment,

$$\frac{40 \text{ B D}^3 d}{\text{W L}^3} = \frac{40 \times \cdot 21}{400 \times 22 \cdot 7} = \cdot 000925 = a.$$

Experiment 3.

58. The next experiments were made with an uniform bar of iron, cast by Messrs. Dowson, 3

[4] The weight of the scale, 8 lbs., ought to have been added.

inches by 1 and $1\frac{1}{2}$ inches, and 6·5 feet between the supports. When this bar was placed on its edge, and loaded in the middle with

150 lbs. the deflexion in the middle was 1 fortieth of an inch.
290 lbs. . . . . 2 do.
360 lbs. . . . . $2\frac{1}{2}$ do.
440 lbs. . . . . 3 do.

The same deflexions were observed in removing the load, and it perfectly regained its natural state. Whence we have,

$$\frac{40 \text{ B D}^3 d}{\text{W L}^3} = \frac{1·5 \times 27 \times 3}{440 \times 274·625} = ·00105 \text{ nearly} = a.$$

Experiment 4.

59. The same piece, with the supports at the same distance, placed flatwise, and loaded in the middle with

180 lbs. the deflexion in the middle was 5 fortieths of an inch.
360 lbs. . . . . 10 do.

The bar restored itself perfectly when the weights were removed, and the trial was repeated with the same results; the load, of 360 lbs., remained upon it ten hours without impairing its elastic force, or increasing the deflexion in the slightest degree.

60. From this and the preceding experiment, the ratio of the breadth and depth to the quantity of deflexion may be compared when the weight is the same. According to the theory of the resistance to flexure (art. 256),

$$d : \frac{1}{\text{B D}^3};$$

and for the weight of 360 lbs. we have

$$\frac{1}{1{\cdot}5 \times 3^3} : \frac{1}{3 \times 1{\cdot}5^3} :: 2\tfrac{1}{2} : \frac{9 \times 2{\cdot}5}{22{\cdot}5} = 10,$$

as it was found to be by experiment.

To find the constant multiplier from the last experiment, we have

$$\frac{40\,\mathrm{B}\,\mathrm{D}^3\,d}{\mathrm{W}\,\mathrm{L}^3} = \frac{3 \times 3{\cdot}375 \times 10}{360 \times 274{\cdot}625} = {\cdot}00102 = a.$$

This value of a does not exactly agree with the one calculated from the first experiment on the same piece; but it is as near as can be expected in a case of this kind; and in a practical point of view it is as near an approach to accuracy as the nature of the subject requires.

Experiment 5.

61. I was desirous of trying the effect of an uniformly distributed load, and my weights, which are cubical pieces of cast iron, all of the same size, and each weighing 10 lbs., are very well adapted for the purpose.

The same piece that was used for the last experiment was laid flatwise upon supports, the supports being 6 feet 6 inches apart, and 18 weights (in all 180 lbs.) were laid along the upper side, just so as to be clear of one another, in the manner shown in fig. 2, Plate I. The deflexion produced by these weights was $\frac{3}{40}$ths of an inch.

A second tier of weights being added, making the

whole weight upon the bar 360 lbs., the deflexion was $\frac{6}{40}$ths of an inch.

62. Hence it appears, that when the weight is uniformly distributed over the length, the deflexion is directly as the weight.

And comparing this with the preceding experiment, it appears, that the deflexion from the weight uniformly distributed over the length, is to the deflexion from the same weight applied in the middle of the length, as 6 is to 10.

The proportion obtained by theoretical investigation is as 5 is to 8; but as $6 : 10 :: 5 : 8\frac{1}{3}$. This small difference arises undoubtedly from error in measuring the deflexions in the experiments.

To compare the value of the constant multiplier by this experiment, the equation

$$\frac{40\,B\,D^3\,d}{\frac{5}{8}\,W\,L^3} = a$$

must be used, whence we find $a = \cdot 00098$.

Experiment 6.

63. This experiment was made upon a piece of iron cast by Messrs. Bramah, of Pimlico, London. It crumbled sooner under the hammer than that used in the preceding experiments, and did not yield quite so readily to the file; it was regular and fine-grained.

The piece was uniform, and $\frac{9}{10}$ths of an inch square; the supports were 3 feet apart, and the weight was applied in the middle of the distance between the supports.

Weight in lbs.	Defl. in in.	Remarks.	Weight in lbs.	Defl. in in.	Remarks.
20	·02		220	·225	
40	·04	When unloaded it returned to its original form; loaded again, the deflexion was the same, and it remained loaded 12 hours without sensible increase, when on being unloaded it was found to have acquired a permanent set of ·02 in. The index was set to nothing, and the weights produced the same deflexions as at first; and it was further loaded as described.	240	·245	
60	·06		260	·27	
80	·08		280	·293	
100	·10		300	·318	When this load had been on 20 minutes, it became ·32 inch.
120	·12		320	·34	
140	·14		340	·365	
160	·162		360	·392	
180	·183		380	·42	
200	·21		400	·445	
			420	·475	
			440	·5	
			460	·532	
			480	·57	which became in an hour ·58.

When the weights were removed, the piece retained a permanent deflexion of ·075 inch; but it was several hours before it returned to that curvature. I did not break the specimen, because I had not weight enough by me for that purpose, neither would it have given a fair measure of the strength of the iron after the trials I have described; but I hope the effect of these trials will make the reader sensible of the necessity of limiting the strain within the range of the elastic force of the material.

According to this experiment,

$$\frac{40\,B D^3 d}{W L^3} = \frac{40 \times \cdot 9^4 \times \cdot 21}{200 \times 27} = \cdot 00102 = a.$$

COMPARISON OF THE PRECEDING EXPERIMENTS.

64. If the mean value of the constant a be taken for the experiments from art. 53 to 63, it is 0·0010446. The number used in calculating the first Table (art. 5, p. 12,) was 0·001, a sufficiently near approximation, with the advantage of much simplicity.

Experiments 7, 8, *and* 9.

65. The next trials were made with specimens formed as shown in fig. 4, Plate I., with the deepest part C D exactly in the middle of the length, and the depth, at C D, 0·975 inch; the depth E A and B F were each half that at C D. The distance of the supported points A B was 3 feet, and the breadth of the bars 0·75 inch. The load was suspended from the point C in the middle of the length, and the deflexion was measured at the same point: the load was increased by 10 lbs. at a time.

Weight acting on the bar.	1st Specimen. Deflex. produced.	2nd Specimen. Deflex. produced.	3rd Specimen. Deflex. produced.
lbs.	in.	in.	in.
40	·052	·065	·052
80	·104	·13	·105
120	·16	·19	·16
160	·215	·25	·21
180	·245	·28	·24
200	·272	·32	·265
500	·84		
540	Broke.		

On the first specimen the load of 180 lbs. remained twelve hours; the deflexion did not sensibly increase, and it returned to its natural form when unloaded; it was again loaded to 200 lbs., which remained upon it two hours; it was then unloaded again, and was found to have taken a permanent set with a deflexion of ·005 inch. The specimen was then loaded again, and the deflexions observed at every 20 lbs.: the deflexion produced by the addition of 20 lbs. was at first ·026, became ·03, ·04, and towards the end of the experiment ·05. When the load had been increased to 360 lbs., in every succeeding addition of 10 lbs. I observed that the deflexion increased by starts of as much as $\frac{1}{100}$th of an inch each, which appeared to be caused by the ends sliding on the supports, at the moment the weight was added; the bar emitted a slight crackling noise, like that produced by bending a piece of tin. There was a small defect in the bar at the place where it broke, which was 4 inches distant from the middle.

When the second specimen was unloaded, immediately, from a weight of 200 lbs. it barely returned to its natural form; but a load of 180 lbs. produced a permanent deflexion of ·005 when it remained upon it fourteen hours.

The load of 200 lbs. remained twenty-one hours upon the third specimen, and when it was unloaded the index returned to zero; therefore this strain was less than would produce a permanent set. The set was nearly ·01 when the load was increased to 210 lbs., and remained upon it ten hours. It was a smoother and better casting than the other specimens.

There did not appear to be any sensible difference in the quality of the iron in these specimens, except that the second specimen was more brittle under the hammer than the other two. They were all fine grained, and yielded easily to the file. They were cast by Messrs. Bramah.

66. I was proceeding with a trial of a piece of the same kind of iron, formed as described in fig. 4, Plate I., when it broke suddenly, at about a foot from the end, at an air bubble The bubble was not apparent on the surface, and yet so near it, that a slight stroke of a hammer would have broken into it. Founders should be very careful to avoid defects of this kind; and beams to sustain great weights should always be proved to a deflexion within their range of elasticity before they are used.

Experiments 10, 11, *and* 12.

67. These trials were made on three pieces of uniform breadth and depth, with the supports 3 feet apart, the load being applied in the middle of the length. The depth ·9 inch, and the breadth the same.

Weight acting on the bar.	1st Specimen. Deflex. produced.	2nd Specimen. Deflex. produced.	3rd Specimen. Deflex. produced.
lbs.	in.	in.	in.
40	·041	·042	·041
80	·082	·09	·08
120	·124	·136	·12
160	·165	·18	·16
180	·185	·202	·18
200	·206		·20

The load of 200 lbs. remained twelve hours on the first specimen, and when it was unloaded the quantity of permanent deflexion was barely sensible; and it was loaded and unloaded again with the same result.

The load of 180 lbs. remained three hours on the second specimen; it had not increased the deflexion, but when the load was removed, it was found that the bar had acquired a permanent set of nearly $\frac{1}{100}$th of an inch.

In the third specimen the bar returned perfectly to its natural form when the load was removed, after being upon it three hours.

Of these specimens the third was the most brittle under the hammer, and the hardest to the file; there was not a sensible difference between the other two; both were soft iron. These specimens were cast by Messrs. Bramah.

68. The chief object in view in the experiments No. 2, 6, 7, 8, 9, 10, and 11, was to determine the strain a square inch of cast iron would bear without permanent alteration, and the extension corresponding to that strain. Calling f this strain in pounds, the experiment 2 gives $f = 15{,}300$ lbs.[5] $= 6 \cdot 8303$ tons, as calculated in art. 143; and the others being

[5] This number is much too great; and the frequent use of it, as an average quantity, has affected the conclusions in the two following articles, and other parts of the work. See note to art. 143, or "Additions," art. 3.—EDITOR.

calculated by the same formula, in experiments 6, 10, and 12, $f=14{,}814$; in experiments 7, 8, and 9, $f=15{,}160$; and in experiment 11, $f=13{,}333$ lbs. The greatest difference amounts to about $\frac{1}{8}$th of the highest value of f; but, in the experiment 2, the load was taken off after remaining only about ten minutes on the bar; in the others it remained for several hours. The former I consider most strictly applicable to practice; and yet it was desirable to show that a force acting a considerable time will produce a permanent set, when the same force could not produce it in a few minutes.

69. In art. 212, it is calculated that the extension produced by the strain of 15,300 lbs. in experiment 2, was $\frac{1}{1204}$ of the length;[6] and by the same mode of calculation the extension in experiment 6 is found to be $\frac{1}{1143}$, in experiment 10, $\frac{1}{1165}$, in experiment 11, $\frac{1}{1170}$, and in experiment 12, $\frac{1}{1200}$. Also, by the equation, art. 127, the extension in experiment 7 is found to be $\frac{1}{1332}$, in experiment 8, $\frac{1}{1132}$, and in experiment 9, $\frac{1}{1367}$.

The difference between the extension in the 8th

---

[6] The extension in experiment 2 has been re-calculated, and found to be the same as here stated, by Professor Leslie, whose mode of calculation is different. See Leslie's Elements of Natural Philosophy, vol. i. p. 240. Edinburgh, 1823.

and 9th experiments is the most considerable; and the mean between these is $\frac{1}{1239}$, which differs very little from $\frac{1}{1204}$, the number used in the rules.

70. A Table of the chief experiments that have been made on the absolute strength of cast iron bars to resist a cross strain, the bars supported at the end, and loaded in the middle.

No.	Description.	Length between the supports in feet.		Dimensions at the strained point in inches.		Weight in lbs. that broke it.	Calculated weight that would destroy elastic force in lbs.	Ratio of the calculated weight to the breaking weight.
		ft.	in.	brdth.	dpth.			
1	Uniform bar	3	0	1	1	756	283	1 : 2·7
2	Ditto	3	0	1	1	735	283	1 : 2·6
3	Ditto	2	6	1	1	1008	340	1 : 2·96
4	Ditto	3	0	1	1	963	283	1 : 3·4
5	Ditto	3	0	1	1	958	283	1 : 3·38
6	Ditto	3	0	1	1	994	283	1 : 3·5
7	Ditto cast from the cupola.	3	0	1	1	864	283	1 : 3·05
8	Parabolic bar cast from the cupola.	3	0	1	1	874	283	1 : 3·08
9	Uniform bar	3	0	1	1	897	283	1 : 3·17
10	Ditto	2	8	1	1	1086	318·75	1 : 3·4
11	Ditto	1	4	1	1	2320	637·5	1 : 3·6
12	Ditto	2	8	2	$\frac{1}{2}$	2185	637·5	1 : 3·42
13	Ditto	1	4	2	$\frac{1}{2}$	4508	1275	1 : 3·53
14	Ditto	2	8	3	$\frac{1}{3}$	3588	956·25	1 : 3·63
15	Ditto	1	4	3	$\frac{1}{3}$	6854	1912·5	1 : 3·58
16	Ditto	2	8	4	$\frac{1}{4}$	3979	1275	1 : 3·12
17	Semi-ellipse	2	8	4	$\frac{1}{4}$	4000	1275	1 : 3·14
18	Parabolic	2	8	4	$\frac{1}{4}$	3860	1275	1 : 3·03
19	Uniform strain in the direction of diagonal	2	8	$\sqrt{2}$	$\sqrt{2}$	851	224·5	1 : 3·79

The two columns on the right-hand side are added to show the relation between the load which permanently destroys a part of the elastic force, and that which breaks the piece. It will be seen that the load which would produce permanent alteration, according to the formula as derived from my experiments, is about $\frac{1}{3}$rd of that which actually broke the specimens; in the worst kind tried, it is $\frac{1}{2 \cdot 6}$ of the breaking weight.

In the preceding Table, the experiments 1, 2, and 3, were made by Mr. Reynolds. No. 1 was twice repeated with the same result. No. 2 is a mean of three experiments.[7] Hence the mean ratio will be about 1 : 2·7. The experiments No. 4, 5, 6, 7, and 8, were made by Mr. Banks;[8] the mean ratio being 1 : 3·3. The rest were made by Mr. George Rennie, and all of the bars of his experiments were cast from the cupola;[9] the mean ratio being 1 : 3·4.

71. Allowing that the preceding experiments are sufficient to fix with considerable certainty the utmost strain that ought to exist in any structure of cast iron, still there is abundant scope for new experimental research; and that which perhaps may be considered of most importance is the effect produced by combining iron of different qualities.

---

[7] Banks on the Power of Machines, p. 89. [8] Idem, p. 90.

[9] Philosophical Transactions for 1818, Part I., or Philosophical Magazine, vol. liii. p. 173.

Through the kindness of Mr. Francis Bramah, I am enabled to begin this inquiry. He has furnished me twelve specimens, of six different kinds of iron; that is, two specimens of each kind. Of these kinds three were run from pig iron from different iron works; one kind was run from old iron, usually termed scrap iron; another kind a mixture of old iron and pig iron in equal parts, and the sixth kind pig iron with an alloy of $\frac{1}{16}$th of copper.

Before I begin to describe the experiments, it will be proper to inform the reader what method I pursued in making them. I knew, from previous trials, that the force which produces a permanent set cannot be determined with that precision which is necessary in comparing iron of different kinds; we can merely observe when it is, and when it is not sensible; and it is most likely that it becomes so by gradations which we cannot trace. It was desirable to ascertain whether a load equivalent to 15,300 lbs. upon a square inch would produce a set or not; and a load of 162 lbs., on the middle of a bar of the size of the specimens, causes that degree of strain: hence, in specimens of the same size, the flexure by this load gives the comparative power of the different kinds, particularly when compared with the quantity of set produced by this or some additional load. But in specimens of different sizes, the comparison is most easily made by calculating the modulus of resilience, or resistance to impulsion, which gives the toughness or relative power of the material to resist a blow. Yet, even then it should be tried what strain will produce permanent alteration, or that which causes fracture, otherwise the comparative goodness of the iron will not be known: I have tried both in each of the varieties of iron.

For all purposes where strength is required, that iron is to be esteemed the best which will bear the greatest degree of flexure without set, and the greatest load. The worst and most brittle pieces of iron have the greatest degree of stiffness; consequently the highest modulus of elasticity; for even the most flexible kind of iron is sufficiently stiff.[10]

In the iron which was taken as a good medium to calculate from, (see experiment, art. 56,) we found

The force that it would bear without permanent alteration	15,300 lbs.
The extension in parts of the length extended .	$\frac{1}{1204}$

[10] I have here followed the principles of comparing materials which were first given in my "Elementary Principles of Carpentry," art. 368-373. The toughness is measured by the same data as in that work, only here a general number of comparison is used instead of making one material a standard of comparison. The term *modulus of resilience,* I have ventured to apply to the number which represents the power of a material to resist an impulsive force; and when I say that one material is tougher than another, it is in consequence of finding this modulus higher for that which is described as the toughest: see arts. 299 to 304, further on.

The modulus of elasticity for a base 1 inch square . . . . 18,400,000 ℔s.
The modulus of resilience . . . . 12·7

These numbers being compared with the results of the experiments now to be described will afford the means of judging both of the qualities of the iron experimented upon, and of the fairness of the mean data I have employed.

OLD PARK IRON.

72. Two specimens run from this kind of pig iron, each 3 feet in length, and smooth, clean, and regular castings, were first put in trial. The section of the bars rectangular; depth 0·65 inch; breadth 1·3 inches; the supports 2·9 feet apart; and the load suspended from the middle.

Weight applied.	Effect on 1st bar.	Effect on 2nd bar.
℔s.	in.	in.
60	bent 0·1	bent 0·1
120	„ 0·2	„ 0·203
162	„ 0·265	„ 0·275
182	„ 0·305 small set.	„ 0·31 {set barely perceptible.
190	„ 0·32 set ·005	„ 0·33 set ·005

The iron was slightly malleable in a cold state; yielded easily to the file. The fracture dark gray with a little metallic lustre; fine grained and compact.

We may consider 162 ℔s. as the greatest load it would bear without impairing its elastic force; and

0·27 is the mean between the flexures produced by this weight; therefore, calculating on these data, we have

The strain it would bear on a square inch without permanent alteration	15,390 lbs.
Extension in length by this strain	$\frac{1}{1152}$
Modulus of elasticity for a base of an inch square	17,744,000 lbs.
Modulus of resilience	13·4
Specific gravity	7·092

The absolute strength to resist fracture was tried by fixing the bar at one end, the load being applied by fixing a scale at the other end, and adding weights till the bar broke. The second bar tried in this manner broke with 184 lbs., the leverage 2 feet; fracture close to the fixed end, metal sound and perfect at the place of the fracture.[11]

Hence, calculating by equation, art. 110, the absolute cohesion of a square inch is 48,200 lbs.,[12] or 3·15 times 15,300 lbs., the strain which has been found incapable of causing permanent set.

Hence I infer, that this iron is superior in toughness, and less stiff than the mean quality.

---

[11] These are circumstances which must have place, otherwise the experiment does not give a fair measure of the strength.

[12] This erroneous conclusion as to the great strength of cast iron, into which Navier, as well as Tredgold, has fallen, arises principally from a supposition that the neutral line remains stationary during the flexure of the body. See "Additions," or Notes to arts. 68 and 143.—Editor.

ADELPHI IRON.

73. The specimens of this iron were clean good castings of the same dimensions as those of Old Park iron. That is, depth 0·65 inch; breadth 1·3 inches; distance between the supports 2·9 feet.

Weight applied.	Effect on 1st bar.	Effect on 2nd bar.
lbs.	in.	in.
60	bent 0·1	bent 0·1
120	,, 0·2	,, 0·205
162	,, 0·26 no set.	,, 0·27 no set.
182	,, 0·3 set ·0075	,, 0·305 set ·005

Comparing this with the preceding experiments on Old Park iron, it is stiffer, and sooner acquires a permanent set. It is also somewhat harder to the file, and more brittle under the hammer. The colour of the fracture was a lighter gray, with less metallic lustre.

Its elasticity is not affected by the load of 162 lbs.; therefore

It will bear upon a square inch without permanent alteration . . . . 15,390 lbs.
And the mean of the two experiments gives the extension . . . . $\frac{1}{1174}$
Modulus of elasticity for a base of 1 square inch 18,067,000 lbs.
Modulus of resilience . . . . 13·1
Specific gravity . . . . 7·07

The second bar, fixed at one end with a leverage of 2 feet, broke with 173 lbs.; the fracture close to the fixed end, and the place of fracture sound and perfect.

According to this experiment, the absolute cohesion is 45,300 lbs. for a square inch, or 2·96 times 15,300 lbs.

A comparison of these trials shows that the difference between Adelphi and Old Park iron is not much, but that the Old Park is superior, particularly in absolute strength; for it required 184 lbs. to break the one, and only 173 lbs. to break the other.

ALFRETON IRON.

74. There was not a sensible difference between the size of these bars and the others. The depth 0·65 inch; breadth 1·3 inches; distance between the supports 2·9 feet.

Weight applied.	Effect on 1st bar.	Effect on 2nd bar.
lbs.	in.	in.
60	bent 0·1	bent 0·1
120	,, 0·2	,, 0·195
162	,, 0·27 no set.	,, 0·28 no set.
183	,, 0·31 small set.	,, 0·325 small set.

This iron differs very little from Old Park, a little more flexible, but very little. It seemed if any thing somewhat harder to the file, but of a less degree of malleability; for instead of extending, it crumbled under the hammer. Fracture scarcely differing from that of Adelphi iron.

These bars bore 162 lbs. without set, and the mean deflexion was ·275. Hence,

The iron would bear upon a square inch without permanent alteration . . . . 15,390 lbs.

Extension in length by this strain	$\frac{1}{1131}$
Modulus of elasticity for a base of 1 inch square	17,406,000 lbs.
Modulus of resilience	13·6
Specific gravity	7·04

The second bar, fixed at one end, broke with 153 lbs., the leverage being 2 feet, the fracture close to the fixed end, and the metal sound and perfect at the place of fracture.

The absolute cohesion, according to this trial, is 40,000 lbs. for a square inch, or 2·63 times the force of 15,300 lbs.

This is a soft species of iron, and may answer extremely well alone, for castings where strength is not required; but it is the weakest iron I have tried, and would most likely be much improved by mixture.

SCRAP IRON.

75. These bars were run from old iron. They were uneven on the surface, indicating that irregularity of shrinkage which has been noticed in the Introduction (page 8). The depth of the bars 0·65 inch; the breadth 1·3 inches; the distance between the supports 2·9 feet.

Weight applied.	Effect on 1st bar.	Effect on 2nd bar.
lbs.	in.	in.
60	bent 0·09	bent 0·09
120	,, 0·18	,, 0·18
162	,, 0·25 no set.	,, 0·255 no set.
180	,, 0·28 no set.	,, 0·285 no set.
190	,, 0·3 small set.	,, 0·3 {set not certain.
210	,, 0·34 set ·005	,, 0·34 set ·004

This iron was very hard to the file, and very brittle, fragments flying off when hammered on the edge, instead of indenting, as the preceding specimens.

The fracture dead or dull light gray; no metallic lustre; not very uniform; fine grained.

These bars showed no sign of permanent set with a load of 180 lbs.; but, to whatever cause this greater range of elastic power may be owing, it certainly would be unsafe to calculate upon it in practice. I shall therefore consider the load of 162 lbs., and a flexure of 0·25 inch, the data to calculate from; accordingly, the

Force of a square inch without permanent alteration	15,390 lbs.
Extension in length by this strain	$\frac{1}{1243}$
Modulus of elasticity for a base of 1 square inch	19,130,000 lbs.
Modulus of resilience	12·4
Specific gravity	7·219

Fixed at one end, with a leverage of 2 feet, the second bar broke with 168 lbs., with one fracture at the fixed end; but the bar flew into several pieces.

This gives the absolute cohesion of a square inch 44,000 lbs., or nearly 2·9 times that strain, which I consider to be the greatest cast iron should have to sustain.

From these elements we may conclude, that a casting of scrap iron will be $\frac{1}{12}$th stiffer than one from Old Park iron; that it has $\frac{1}{12}$th less power to resist a body in motion, and that it is less strong in the ratio of 168 to 184.

MIXTURE OF OLD PARK AND GOOD OLD IRON IN EQUAL PARTS.

76. The castings run from this mixture were even and clean; such as indicate a perfect union of the materials. The depth 0·65 inch; the breadth 1·3 inches; and the distance between the supports 2·9 feet.

Weight applied.	Effect on 1st bar.	Effect on 2nd bar.
lbs.	in.	in.
72	bent 0·1	bent 0·1
140	„ 0·2	„ 0·2
162	„ 0·24 no set.	„ 0·245 no set.
182	„ 0·27 no set.	„ 0·28 no set.
202	„ 0·3 small set.	„ 0·31 small set.
220		„ 0·34 set ·005
300		„ 0·475 set ·03

This iron was rather hard to the file; it indented with the hammer, but was rather short and crumbling.

Fracture a lighter gray, and more dull than Old Park iron; very compact, even, and fine grained.

The bars did not set with a load of 182 lbs., therefore the load of 162 lbs. is sufficiently within the limit; the flexure with that load is ·245, consequently we may state its properties as under:

Force on a square inch that does not produce permanent alteration	15,390 lbs.
Extension under this strain	$\frac{1}{1268}$
Modulus of elasticity for a base of 1 square inch	19,514,000 lbs.
Modulus of resilience	12·1
Specific gravity	7·104

When the second bar was fixed at one end, it broke with 174 lbs., acting with a leverage of 2 feet; fracture close to the fixed end. Therefore the absolute cohesion of a square inch is 45,600 lbs., or very nearly three times the strain of 15,300 lbs.

In this mixture there is clearly too great a proportion of old iron; it is rather inferior to the quality of our mean specimen (art. 56) About one of old iron to two of the Old Park pig iron would be a better proportion. It is worthy of remark, that the absolute strength is nearly the mean of the two kinds which form the mixture, and so is the specific gravity.

ALLOY OF PIG IRON SIXTEEN PARTS, COPPER ONE PART.

77. It has been said that iron is much improved by a small proportion of copper; it was desirable, therefore, to ascertain its effect, and the advantage, if any, of employing it. The breadth of the specimens 1·25 inches; the depth ·675 inch; the distance between the supports 2·9 feet. The load which ought not to produce permanent alteration, about 167 lbs.

Weight applied.	Effect on 1st bar.	Effect on 2nd bar.
lbs.	in.	in.
60	bent 0·1	bent 0·1
122	,, 0·2	,, 0·2
167	,, 0·275 no set.	,, 0·265 no set.
180	,, 0·3 no set.	,, 0·29 no set.
203	,, 0·34 set ·003	,, 0·325 set ·002
300	,, 0·5	

These bars yielded freely to the file, but were short and crumbling under the hammer. I expected to have found them more ductile. The fracture dark gray, fine grained, and more compact than Old Park iron; with less metallic lustre.

The load of 167 lbs. did not produce any degree of set, the mean flexure by this load 0·27; and assuming this to be as great a load as it should bear in practice, we have,

Force on a square inch that does not produce permanent alteration	15,300 lbs.
Extension under this strain	$\frac{1}{1106}$
Modulus of elasticity for a base of 1 inch square	16,921,000 lbs.
Modulus of resilience	13·8
Specific gravity	7·13

To try the absolute strength, the second bar was fixed at one end, and the scale suspended from the other end; weights were then added till the bar broke: the fracture took place close to the fixed end, and it required 194 lbs. to break the bar.

According to this experiment, the cohesive force of a square inch is 52,000 lbs., or 3·4 times the strain that will not give permanent alteration.

It appears that copper increases both the strength and extensibility of iron.

EXPERIMENTS ON THE RESISTANCE TO TENSION.

78. According to an experiment made by Muschenbroëk, a parallelopipedon, of which the side

was ·17 of a Rhinland inch, broke with 1930 lbs.;[13] and since the Rhinland foot is 1·03 English feet, and the pound contains 7038 grains, this experiment gives 63,286 lbs. for the weight that would tear asunder a square inch, when reduced to English weights and measures.

79. An experiment made by Capt. S. Brown is thus described: "A bar of cast iron, Welsh pig, $1\frac{1}{4}$ inch square, 3 feet 6 inches long, required a strain of 11 tons 7 cwt. (25,424 lbs.) to tear it asunder: broke exactly transverse, without being reduced in any part; quite cold when broken; particles fine, dark blueish gray colour."[14]

Capt. Brown's machine for trying such experiments being constructed on the principle of a weigh-bridge, Mr. Barlow is of opinion it may show less than its real force; it also may be remarked, that to obtain the real force of cohesion, the resultant of the straining force should coincide exactly with the axis of the piece, for so small a deviation in this respect as $\frac{1}{6}$th of the breadth would reduce the strength one half.

From this experiment it appears that 16,265 lbs. will tear asunder a square inch of cast iron.

80. In some experiments made by Mr. G. Rennie, it is obvious, from the description of the apparatus,

---

[13] Muschenbroëk's Introd. ad Phil. Nat. vol. i. p. 417. 1762.

[14] Essay on the Strength of Timber, &c., by Mr. Barlow, 1817, p. 235. Reprinted, with additions, 1837.

that the strain on the section of fracture would not be equal; and, therefore, that the straining force would be less than the cohesion of the section. The specimens were 6 inches long, and ¼th of an inch square at the section of fracture. A bar cast horizontally required a force of 1166 lbs to tear it asunder. A bar cast vertically required a force of 1218 lbs. to tear it asunder.[15]

	Per square inch.
In the horizontal casting the force was equal to	18,656 lbs.
And in vertical casting	19,488 „

EXPERIMENTS ON THE RESISTANCE TO COMPRESSION IN SHORT LENGTHS.

81. The power of cast iron to resist compression was formerly much over-rated. Mr. Wilson estimated the power necessary to crush a cubic inch of cast iron at 1000 tons = 2,240,000 lbs.; and in describing an experiment by Mr. William Reynolds, of Ketley, in Shropshire, a cube of ¼th of an inch of cast iron, of the quality called gun-metal, was said to require 448,000 lbs. to crush it.[16] But Mr. Telford, for whom the experiments were made, was so kind as to communicate the correct results of the

---

[15] Phil. Transactions for 1818, Part I., or Philosophical Magazine, vol. liii. p. 167.

[16] Edin. Encyclo. art. Bridge, p. 544; or Nicholson's Journal, vol. xxxv. p. 4. 1813.

experiments made by Mr. Reynolds; and it appears that

	Per square inch.
A cube of $\frac{1}{4}$th of an inch of soft gray metal was crushed by 80 cwt.	= 143,360 lbs.
Ditto of the kind of cast iron called gun-metal was crushed by 200 cwt.	= 350,400 lbs.

82. Such was the state of our knowledge on this important subject, when Mr. G. Rennie communicated a valuable series of experiments to the Royal Society, which were published in the first part of their Transactions for 1818.

Mr. Rennie's Experiments on cubes from the middle of a large block; specific gravity 7·033:

	in.		lbs.		Force per sq. in. in lbs.
Side of cube	$\frac{1}{8}$	was crushed by	1,454,	highest result	= 93,056
Ditto	$\frac{1}{8}$	do.	1,416,	lowest ditto	= 74,624
Ditto	$\frac{1}{4}$	do.	10,561,	highest ditto	= 168,976
Ditto	$\frac{1}{4}$	do.	9,020,	lowest ditto	= 144,320

On cubes from horizontal castings, specific gravity 7·113

	in.		lbs.		lbs. per sq. in.
Side of cube	$\frac{1}{4}$	was crushed by	10,720,	highest result	= 171,520
Ditto	$\frac{1}{4}$	do.	8,699,	lowest ditto	= 139,184

On cubes from vertical castings, specific gravity[17] 7·074

	in.		lbs.		lbs. per sq. in.
Side of cube	$\frac{1}{4}$	was crushed by	12,665,	highest result	= 202,640
Ditto	$\frac{1}{4}$	do.	9,844,	lowest ditto	= 157,540

[17] It is singular that the specific gravity of the vertical castings should be less than that of the horizontal ones.

On pieces of different lengths.

			in.		lbs.	lbs. per sq. in.
Area $\frac{1}{8}$	× $\frac{1}{8}$	length	$\frac{3}{4}$	was crushed by	1,743	= 111,552
Ditto $\frac{1}{8}$	× $\frac{1}{8}$	,,	1	do.	1,439	= 92,096
Ditto $\frac{1}{4}$	× $\frac{1}{4}$	,,	$\frac{1}{2}$	do.	9,374	= 149,984
Ditto $\frac{1}{4}$	× $\frac{1}{4}$	,,	1	do.	6,321	= 101,136

These experiments were on too small a scale to allow of that precision in adjustment which theory shows to be essential in such experiments; therefore there still remains much to be done by future experimentalists. It does not appear, within the limits of these experiments, that an increase of length had any sensible effect on the result.

I have selected the highest and lowest results, and such of the single trials that were made under the greatest difference of length; in all Mr. Rennie made thirty-nine trials on the resistance of cast iron to compression.[18]

EXPERIMENTS ON THE RESISTANCE TO COMPRESSION OF PIECES OF CONSIDERABLE LENGTH.

83. The only experiments of this kind that I know of were made by Mr. Reynolds, and are described as follows in Mr. Banks's work on the 'Power of Machines,' p. 89.

" Experiments on the strength of cast iron, tried at Ketley, in March, 1795. The different bars were

[18] Philosophical Transactions for 1818, Part I., or Philosophical Magazine, vol. liii. pp. 164, 165.

all cast at one time out of the same air furnace, and the iron was very soft, so as to cut or file easily.

"Exp. 1. Two bars of iron, 1 inch square, and exactly 3 feet long, were placed upon an horizontal bar, so as to meet in a cap at the top, from which was suspended a scale; these bars made each an angle of 45° with the base plate, and of consequence formed an angle of 90° at the top: from this cap was suspended a weight of 7 tons (15,680 lbs.), which was left for sixteen hours, when the bars were a little bent, and but very little.

"Exp. 2. Two more bars of the same length and thickness were placed in a similar manner, making an angle of 22½° with the base plate; these bore 4 tons (8960 lbs.) upon the scale: a little more broke one of them which was observed to be a little crooked when first put up."

84. By the principles of statics,[19]

$$2 \sin. 45^\circ : \text{Rad.} :: 15{,}680 \text{ lbs.} : 11{,}087 \text{ lbs.}$$

equal the pressure in the direction of either bar in the first experiment. And,

$$2 \sin. 22\tfrac{1}{2}^\circ : \text{Rad.} :: 8960 \text{ lbs.} : 11{,}709 \text{ lbs.}$$

the pressure in the direction of either bar in the second experiment.

If we consider the direction of the force to have been exactly in the axis in these trials, then, according to the equation, art. 288, the greatest

[19] Gregory's Mechanics, vol. i. art. 48.

strain in the direction of one of these bars should not have exceeded 5840 lbs.; but if the direction of the pressure was at the distance of half the depth from the axis, which it is very probable it would be, the greatest strain in actual construction should not have exceeded 2720 lbs. See art. 287.

EXPERIMENTS ON THE RESISTANCE TO TWISTING.

85. Table of the principal experiments of the strength of cast iron to resist a twisting strain.

No.	Description.	Leverage.	Length.	Side or diameter in inches.	Weight in lbs. that broke the piece.	Calculated resistance without destroying the elastic force.	Ratio of the calculated resistance to the breaking wt.
1	Bar placed vertically, fast at one end & twisted by a wheel at the other.	ft. in. 1 0	not given.	1 × 1	631	150	1 : 4·2
2	Cylinder fixed at one end, twisted by a lever at the other.	14 2	in. $2\frac{3}{4}$	in. 2	250	73·7	1 : 3·39
3	Ditto.	14 2	$3\frac{1}{4}$	$2\frac{1}{4}$	384	111	1 : 3·46
4	Ditto.	14 2	3	$2\frac{1}{2}$	408	140	1 : 2·9
5	Ditto.	14 2	3	$2\frac{3}{4}$	700	184	1 : 3·8
6	Ditto.	14 2	4	$3\frac{1}{4}$	1170	309	1 : 3·78
7	Ditto.	14 2	5	$3\frac{1}{2}$	1240	402	1 : 3·08
8	Ditto.	14 2	5	$3\frac{3}{4}$	1662	481	1 : 3·45
9	Ditto.	14 2	5	4	1938	580	1 : 3·34
10	Ditto.	14 2	6	$4\frac{1}{4}$	2158	713	1 : 3·02

The experiment No. 1 was made by Mr. Banks.[20] The others were made by Mr. Dunlop, of Glasgow.

[20] "Power of Machines."

Nos. 4 and 7 were faulty specimens.[21] Some experiments on a very small scale were made by Mr. George Rennie, but they are not inserted here, because they were not sufficiently described to admit of comparison.[22]

I am indebted to Messrs. Bramah for a description of some new and interesting experiments on torsion, which they had made in order to ascertain what degree of confidence they might place in the theoretical and experimental deductions of writers on this subject. They were also desirous of knowing the effect of a small portion of copper on the quality of cast iron.

I have given a tabular form to the results of these experiments, in order that they may be more easily compared; and I have added two columns to the Table, to show how these experiments agree with the rules of this work.

The bars were firmly fixed at one end, in a horizontal position, and to the other end the straining force was applied, acting with a leverage of 3 feet. To prevent the effect of lateral stress, the bar rested loosely upon a support at the end to which the straining force was applied.

---

[21] Dr. Thomson's Annals of Philosophy, vol. xiii. p. 200-203.

[22] Philosophical Magazine, vol. liii. p. 168.

No. of experiment.	Description of iron.	Length of bar.	Side of bar.	Weight applied acting with 3 feet leverage.	Effect of weight or angle of torsion.	Calculated angle of torsion.	Ratio of the force which would not produce set to the breaking weight.
		ft.	in.	lbs.	deg.	deg.	
1	Square bar of an alloy of 16 parts iron to one of copper.	1	$1\frac{1}{16}$	166	$7\frac{1}{2}$	4·25	
				215	broke		1 : 3·6
2	Square bar of the same kind as number 1.	2	$1\frac{1}{16}$	111	$6\frac{1}{2}$	5·7	
				213	17	10·9	
				213	broke		1 : 3·5
3	Square bar a mixture of equal parts of Adelphi, Alfreton, & old iron.	1	$1\frac{1}{16}$	217	14	5·6	
				330	broke		1 : 5·5
4	Bar same kind as number 3.	1	$1\frac{1}{16}$	166	$7\frac{1}{2}$	4·25	
				310	broke		1 : 5·16
5	Bar same kind as number 3.	2	$1\frac{1}{16}$	164	$12\frac{1}{2}$	8·4	
				213	18	10·9	
				280	28	14·3	
					Broke by slipping one of the weights.		
6	Square bar of cast iron.	1	1	237	broke		1 : 4·72
7	Square bar same kind as number 6.	2	1	218	broke		1 : 4·35

The comparison between our rule (Equation iv. art. 265), and the force that broke these specimens, which is given in the last column, is very satisfactory, and very nearly agrees with former experiments on this strain as shown in the first Table of this article.

The observed angle of torsion is very irregular,

and in all these experiments it greatly exceeds the angle calculated by Equation xiv. art. 272. But it will be remarked, that the angle was measured after the strain was far beyond that degree where it is known that flexure increases more rapidly than the load; and no allowance was made for the compression at the fixed points. M. Duleau, in his experiments on wrought iron, (see Sect. VI. art. 94 of this Essay,) allowed for the latter source of error by taking, as the measure of torsion, the angle through which the bar returned when the weight was taken off;[23] and the formula applied to his experiments gives an error in excess,—here it is in defect: I shall, therefore, not endeavour to make the rules agree with either set of experiments, because I know that the flexure will be too great in Messrs. Bramah's experiment; and assuming this to be so, the rules will be nearly true; whereas, if M. Duleau's turn out to be most correct, it will only cause shafts to be made a small degree stronger than necessary.

EXPERIMENTS ON THE EFFECT OF IMPULSIVE FORCE.

86. The height from which a weight might fall upon a piece of cast iron without destroying its elastic force was calculated by Equation v. art. 306, for the specimens of ·9 inch square, used in the

---

[23] Essai sur la Résistance du Fer Forgé, p. 49.

preceding experiments (art. 67). Repeated trials with that height of fall were made without producing a sensible effect. I then let the weight fall from double the calculated height, and every repetition of the blow added about $\frac{1}{100}$th of an inch to the curvature of the bar. I could not measure the effect of each trial very correctly, but a few trials rendered the bar so much curved as to be easily seen. I hope, at some future time, to be able to resume these experiments with an apparatus for measuring correctly the degree of permanent set.[24] See art. 313-350, where practical rules will be found.

TO DISTINGUISH THE PROPERTIES OF CAST IRON BY THE FRACTURE.

87. I shall close this section with a few remarks on the aspect of cast iron recently fractured, with a view to distinguish its properties.

There are two characters by which some judgment may be formed; these are the colour and the lustre of the fractured surface.

The colour of cast iron is various shades of gray; sometimes approaching to dull white, sometimes dark iron gray with specks of black gray.

The lustre of cast iron differs in kind and in

---

[24] This hope of the Author had not been realized when Practical Science was unfortunately deprived, by his death, of one of its most able supporters.—ED.

degree. It is sometimes metallic, for example, like minute particles of fresh cut lead distributed over the fracture; and its degree, in this case, depends on the number and size of the bright parts. But in some kinds, the lustre seems to be given by facets of crystals disposed in rays. I will call this lustre, crystalline.

In very tough iron the colour of the fracture is uniform dark iron gray, the texture fibrous, with an abundance of metallic lustre. If the colour be the same, but with less lustre, the iron will be soft but more crumbling, and break with less force. If the surface be without lustre, and the colour dark and mottled, the iron will be found the weakest of the soft kinds of iron.

Again, if the colour be of a lighter gray with abundance of metallic lustre, the iron will be hard and tenacious; such iron is always very stiff. But if there be little metallic lustre with a light colour, the iron will be hard and brittle; it is very much so when the fracture is dull white; but in the extreme degrees of hardness, the surface of the fracture is grayish white and radiated with a crystalline lustre.

There may be some exceptions to these maxims, but I hope they will nevertheless be of great use to those engaged in a business which is every day becoming more important.

SECTION VI.

EXPERIMENTS ON MALLEABLE IRON AND OTHER METALS.

EXPERIMENTS ON THE RESISTANCE OF MALLEABLE IRON TO FLEXURE.

88. There have been a greater number of experiments made on malleable iron than on any other metal; but those on the lateral strength are chiefly by foreign experimentalists. From those of Duleau I shall select a few for the purpose of comparison; but in the first place I propose to describe some of my own trials.

The following experiments were made on bars of English and of Swedish iron; the bars were supported at the ends, and the weight applied in the middle between the supports; the length of each bar was exactly 6 feet, and the distance between the supports $66\frac{1}{2}$ inches.

English Iron.

Kind of bar and dimensions.	Weight of 6 feet in length.	Deflexion with			Weight of modulus of elasticity for a base of 1 sq. inch.
		58 lbs.	114 lbs.	170 lbs.	
	lbs.	inch.	inch.	inch.	lbs.
Bar $1\frac{1}{4}$ inch square	33	·0625	·1	·1875	27,240,000
Bar $1\frac{1}{8}$,,	25	·125	·25	·375	20,830,000
Bar 1 ,,	20	·15	·32	·5	24,990,000
Round bar $1\frac{1}{4}$ in. dia.	24	·125	·25	·375	23,154,000
Round bar 1 ,,	17	·25	·5	·8	26,500,000
Mean weight of modulus 24,542,800 lbs.					

Swedish Iron.

Kind of bar and dimensions.	Weight of 6 feet in length.	Deflexion with			Weight of modulus of elasticity for a base of 1 sq. inch.
		58 lbs.	114 lbs.	170 lbs.	
	lbs.	inch.	inch.	inch.	lbs.
Bar 1·2 inch square	32	·0625	·125	·19	32,000,000
Bar $1\frac{1}{8}$,,	27	·08	·161	·25	31,245,000
Bar 1 ,,	33	·125	·25	·375	33,328,000
Mean weight of modulus 32,191,000 lbs.					

The bars of Swedish iron varied in dimensions considerably; the dimension in the first column was taken at the point of greatest strain in each bar. The apparently superior stiffness of the Swedish iron is partly to be attributed to this cause; but it is in a greater degree owing to the mode of manufacture, which gives more density as well as elastic force to the iron. If the English iron had been formed under the hammer, in the same manner, it would have been perhaps equally dense and strong, and as fit for the nicer purposes of smiths' work as the Swedish. All these specimens were tried in the same state as the bars are sent from the iron works; the trials were made in July, 1814.

89. The objects of my next experiments on malleable iron were, to determine the force that would produce permanent alteration; the effect of heating iron so as to give it uniform density; and the effect of temperature on its cohesive power. For this purpose, Mr. Barrow, of East Street, selected for me a bar of what he esteemed good iron, bearing the mark Penydarra. A piece 38 inches long, weighing 10·4 lbs., was cut off this bar; its section did not sensibly differ from 1 inch square. With the supports 3 feet apart, and the weight applied in the middle, the following results were obtained.

Weight.	Deflexion in the middle with the bar as obtained from the iron works.	Deflexion in the middle after the bar had been uniformly heated and slowly cooled.
lbs.	inch.	inch.
126	·05	·059
252	·10	·117
310	·12	·145
330	·13	·154

In both states it bore the weight of 330 lbs. without sensible effect, though it was let down upon it, and relieved several times; but in either state an addition of 20 lbs. rendered the set perceptible; in the softened bar it appeared to be sensible when only 10 lbs. had been added.

Hence, by art. 110 we have the force that could be resisted; without permanent alteration 17,820 lbs. per square inch: by art. 121 the extension, in the softened state, is $\frac{1}{1400}$ of its length; and by art. 105 the modulus of elasticity is 24,920,000 lbs. for a base of an inch square. The modulus before being softened is 29,500,000 lbs.

90. To try the effect of heat in decreasing the cohesion of malleable iron, I heated it to 212° of Fahrenheit, having previously got the machine ready, so that a weight of 300 lbs. could be instantly let down upon the bar as soon as it was put in, and the index adjusted to one of the divisions of the scale. These operations having been effected in a close and warm room, with as little loss of heat as possible, the window was thrown open, and the effect of cooling observed. The deflexion decreased as the bar cooled, but it was allowed to remain nearly two hours, in order to be perfectly cooled down to the temperature of the room, or 60° Each division on the scale of the index is $\frac{1}{100}$th of an inch, and as nearly as I could determine, with the assistance of a magnifier, the deflexion had decreased three-fourths of one of the divisions; and it returned through fourteen divisions when the load was removed; therefore we may conclude, that by an elevation of temperature equal to $212-60=152$ degrees, iron loses about a 20th part of its cohesive force, or a 3040th part for each degree.

M. DULEAU'S EXPERIMENTS.[1]

91. The most part of the experiments of M. Duleau were made with malleable iron of Perigord; some of the specimens were hammered to make them regular, others were put in trial in the state they are sent from the iron works; the former are distinguished from the latter by an *h* added to the number of the experiment; and these numbers are the same as in M. Duleau's work. The experiments are divided into two classes: in the first the elasticity was observed to be impaired by the action of the load; in the second it was not. The specimens were supported at the ends, and the load suspended from the middle of the length. The dimensions of the pieces are in the original measures, as well as the weights; but the deductions are in our own measures and weights. All the experiments I have selected are on Perigord iron.

Number of experiments.		Distance between the supports.	Breadth.	Depth.	Depression in the middle.	Weight producing it.	Extension in parts of length.
		Millim.	Millim.	Millim.	Millim.	Kilogrammes.	
1st class	15	2000·	45·	12·	54·	45	·000972
	17^h	2000·	40·	11·5	52·5	25	·000906
	36^h	3000·	60·	20·	33·	50	·000441
2nd class	21^h	2000·	11·5	40·	15·03	90	·000902
	22	3000·	77·	14·	72·	50	·000672
	29^h	3000·	15·	25·	70·	50	·001167

[1] Taken from his Essai Théorique et Expérimental sur la Résistance du Fer Forgé. 4to. Paris, 1820.

The last column shows the extension of an unit of length by the strain as calculated by M. Duleau; my formula, art. 121, gives the same results. The extension that malleable iron will bear without permanent alteration is $\frac{1}{1400}$ = ·000714, according to my experiment; but in M. Duleau's experiment, No. 36[h], the extension of ·000441 produced a permanent set, while in No. 29[h] the extension was ·001167 without producing a set: this is a considerable irregularity, but such as may be expected in experiments on such long heavy specimens of small depth. In all such experiments, the effect of the weight of the piece should be observed. It is also essential that the points of support should be perfectly solid and firm, or that the flexure should be measured from a point, of which the position is invariable in respect to the points of support.

The mean weight of the modulus of elasticity, as determined by the above experiments, is 28,000,000 lbs. for a base of 1 inch square. Experiment No. 22 gives the highest, being 31,864,000 lbs.; and No. 17, the lowest, being 22,974,000 lbs.; therefore it appears that the elastic force of Perigord iron is not greatly different from English iron.

M. Duleau concludes that a bar of malleable iron may be safely strained till the extension at the point of greatest strain is equal to $\frac{1}{3333}$ of its original length without losing its elasticity; and that the load upon a square inch which produces this extension is 8540 lbs. In many of his own experiments the extension was three times this without permanent loss of elasticity.

It has been my object to fix the limit which will produce permanent alteration of elasticity in a good material; to say, that beyond this strain you must not go, but approach it as nearly as your own judgment shall direct, when you are certain that you have assigned the greatest possible load it will be exposed to. Where a great strain is to be sustained, a good material is most suitable and most economical; to a defective material no rules whatever will apply; for who can measure the effect of a flaw in malleable iron, an air bubble in cast iron, a vent in a stone, or of knots and rottenness in timber? But the presence of most of these defects can be ascertained by inspection of the material itself; and since the greatest strain is at the surface of a beam or bar, the defects which impair the strength in the greatest degree are always most apparent.

Experiments on the flexure of malleable iron have also been made by Rondelet,[2] Aubry, and Navier,[3] which accord with the theoretical principles developed in this Essay.

---

[2] Traité de l'Art de Bâtir, tome iv. p. 509 and 514. 4to. 1814.

[3] Gauthey's Construction des Ponts, tome ii. p. 151. 4to. 1813.

EXPERIMENTS ON THE RESISTANCE TO TENSION.

92. The experiments on the absolute resistance of malleable iron to tension are very numerous: in many experiments it has been found above 80,000 ℔s. per square inch, and in very few under 50,000 ℔s., indeed in none where the iron was not defective. About 60,000 ℔s. seems to be the average force of good iron; and according to this estimate, the force that would produce permanent alteration is to that which would pull a bar asunder as 17,800 : 60,000, or nearly as 1 : 3·37. Hence we see, that on whatever principle it was that Emerson[4] concluded a material should not be put to bear more than a third or a fourth of the weight that would break it, the maxim is agreeable with the laws of resistance.

Experiments on the absolute strength of malleable iron have been made by Muschenbroëk,[5] Buffon,[6] Emerson,[7] Perronet,[8] Soufflot,[9] Sickingen,[10] Rondelet,[11] Telford,[12] Brown,[13] and Rennie.[14] Those

---

[4] Mechanics, 4to edit. p. 116. 1758.

[5] Introd. ad Phil. Nat. i. p. 426. 4to. 1762.

[6] Gauthey's Construction des Ponts, ii. pp. 153, 154.

[7] Mechanics, p. 116. 4to edit. 1758.

[8] Gauthey's Construction des Ponts, ii. pp. 153, 154.

[9] Rondelet's L'Art de Batir, iv. pp. 499, 500. 4to. 1814.

[10] Annales de Chimie, xxv. p. 9.

[11] Rondelet's L'Art de Bâtir, iv. pp. 499, 500. 4to. 1814.

[12] Barlow's Essay on Strength of Timber, &c. pp. 221-237. 1817. Reprinted, 1837.

[13] Ibid.

[14] Philosophical Magazine, liii. p. 167. 1819.

by Messrs. Telford and Brown were made on the largest scale; and are minutely described in Professor Barlow's Essay, to which I must refer the reader.

EXPERIMENTS ON THE RESISTANCE TO COMPRESSION.

93. Very few experiments have been made on this species of resistance, and from some circumstances in such experiments requiring attention which the authors of them do not appear to have been aware of, we can make no use of them in illustrating our theoretical principles, unless it be to show that when we consider the direction of the force to nearly coincide with one of the surfaces of the bar, we shall always be calculating on safe data; and from the nature of practical cases in general, we can scarcely think of employing a less excess of force than is given by this rule.

On pieces of considerable length experiments have been made by Navier, Rondelet, and Duleau; and the force necessary to crush short specimens has also been ascertained by Rondelet.

Rondelet employed cubical specimens, the sides of the cubes varying from 6 to $10\frac{1}{2}$ and 12 lines; and cylinders of 6, 8, and 12 lines in diameter, the height being the same as the diameter in each cylinder. The mean resistance of the cubes was equivalent to 512 livres on a square line; the mean resistance of the cylinders 515 livres per square line:

512 livres on a square line is 70,000 lbs. on a square inch in our weights and measures. The force necessary to crush the specimens was in proportion to the area; when the area was increased four times, this ratio did not differ from the result of the experiment so much as a fiftieth part.[15]

He observed in experiments on bars of different lengths, that when the height exceeded three times the diameter, the iron yielded by bending in the manner of a long column. Rondelet's experiments on longer specimens are not sufficiently detailed.[16]

Navier's experiments were made on long bars, and show the force that broke them; whether the flexure was sudden or gradual is not stated.[17]

A bar of any material, in which the stress is very accurately adjusted in the direction of the axis, will bear a considerable load without apparent flexure, but the load is in unstable equilibrium, so much so indeed, that in a bar where the least dimension of the section is small in respect to the length, the

---

[15] There does not appear to be an abrupt change in the crushing of wrought iron to enable an experimenter to draw any very definite conclusions of this kind. According to my observations, wrought iron becomes slightly flattened or shortened with from 9 to 10 tons per square inch; with double that weight it is permanently reduced in length about $\frac{1}{62}$, and with three times that weight about $\frac{1}{16}$th of its length. (Philosophical Transactions, Part II., 1840, p. 422.)—Editor.

[16] Traité de l'Art de Bâtir, iv. pp. 521, 522.

[17] Gauthey's Construction des Ponts, ii. p. 152.

slightest lateral force would cause the bar to bend suddenly and break under the load. In such a case, it is not so much owing to the magnitude of the force that fracture is produced, as the momentum it acquires before the bar attains that degree of flexure which is necessary to oppose it. The reader will find this view of the subject to be agreeable to experience, particularly in flexible materials; in fact, I do not think any one can be aware of the danger of over-loading a column who has never observed an experiment of this kind.

M. Duleau found that a bar of malleable iron 11·8 feet long, and 1·21 inches square (31 millimétres), doubled under a load of 4400 lbs. (2000 kilogrammes). Another specimen about 11·8 feet long, the breadth 2·38 inches, and the depth 0·8 inch, doubled under a load of 2640 lbs.: this piece did not become sensibly bent before it doubled.[18] In the last experiment, our rule (Equa. xv. art. 288,) gives 876 lbs. as the greatest load the bar ought to sustain in practice; which is about one-third of the weight that doubled the piece; a similar result obtains in other cases.

EXPERIMENTS ON THE RESISTANCE TO TORSION.

94. Mr. Rennie made some experiments on the resistance of malleable iron to torsion. The weight

---

[18] Essai Theorique et Experimental sur la Résistance du Fer Forgé, p. 26-37.

acted with a lever of 2 feet, and the specimens were ¼th of an inch square; the strain was applied close to the fixed end:

	lbs.	oz.
English iron, wrought, was wrenched asunder by	10	2
Swedish iron, wrought, ... by	9	8 [19]

If we could suppose the pieces so fitted that the distance between the centres of action, of the force, and the fixing apparatus, was equal to the diameter of the specimen; then our formula gives 1·315 lbs. as the force that such a bar would resist without permanent change: this is only about ⅛th of the force that produced fracture. A like irregularity occurs in his experiments on the torsion of cast iron, which may very likely be in consequence of the strain not being applied exactly as I have supposed it to be.

The experiments on the resistance of malleable iron to torsion made by M. Duleau were all directed to determining its stiffness. The bars were fixed at one end in a horizontal position, and the force was applied to a wheel or large pulley fixed on the other end. In order to prevent lateral strain, the end to which the wheel was fixed reposed freely upon a support. It was found that the bars yielded a little at the fixed points; the permanent alteration produced by this yielding was allowed for by deducting the angle of set from the angle observed.[20]

---

[19] Philosophical Magazine, vol. liii. p. 168.

[20] Essai sur la Resistance du Fer Forge, p. 50-53.

Nature of the specimens.	Length of the part twisted.	Sides of diameter.	Angle of torsion with a wt. of 10 kilogrammes (22 lbs.) with leverage of 320 millimétres (1·22 feet).	Angle of torsion as calculated.
	Millimétres.	Millimétres.	degrees.	degrees.
Round iron, English, marked Dowlais, as from the iron works; hot short.	2400 (7·9 ft.)	19·83 (·78 in.)	4	10·4
Round iron, Perigord, as from the iron works.	2890 (9·5 ft.)	23·03 (·91 in.)	3	7
Square iron, English, marked *C* 2, hot short.	4120 (13·5 ft.)	20 × 20 (·79 in.)	6½	10
Square iron, Perigord, as from the iron works.	2520 (8·3 ft.)	20·35 × 20·35 (·8 in.)	3·08	5·8
Flat iron, English.	2910 (9·6 ft.)	34 × 8·56 (1·32 × ·337 in.)	11·4	13·9

The last column shows the angle calculated by the formula, (Equa. xiii. and xiv. art. 272). There is a considerable error in excess according to these experiments; see art. 85.

EXPERIMENTS ON VARIOUS METALS.

EXPERIMENTS ON STEEL.

95. The modulus of elasticity of steel was first determined by Dr. Young from the vibration of a tuning-fork; the height of the modulus found by this method was 8,530,000 feet;[21] hence the weight

[21] Lectures on Natural Philosophy, vol. ii. p 86.

of the modulus for a base of an inch square will be 29,000,000 lbs.

M. Duleau has made some experiments on the flexure of steel bars when loaded in the middle and supported at the ends; in all he has described twelve experiments;[22] from these I will take four at random.

Description of specimens.	Distance between the supports.	Breadth.	Depth.	Depression with 10 kilogrammes.	Weight of modulus of elasticity in lbs. for a base an inch square.
	Millim.	Millim.	Millim.	Millim.	English lbs.
English cast steel, marked HUNTSMAN, perfectly regular, untempered, but brittle.	980	13·3	5·9	32·05	34,000,000
German steel (of cementation), marked FORTSMAN, and 3 deer heads, used for razors, dimensions irregular.	680	14·5	7·8	8	20,263,000
Same kind of steel.	1845	28·5	21·9	2·6	29,000,000
Ditto.	1350	52	26·6	0·5	17,880,000
Mean for German steel 22,381,000 lbs.					

EXPERIMENTS ON GUN-METAL.

96. A cast bar of the alloy of copper and tin, commonly called gun-metal, of the specific gravity 8·152, was filed true and regular; its depth was 0·5 inch, and its breadth 0·7 inch; it was supported

[22] Essai sur la Résistance du Fer Forge, p. 38.

at the ends, the distance between the supports being 12 inches; and the scale was suspended from the middle.

Load	Deflexion	Remarks
19 lbs. bent the bar	0·01 inch.	
38	0·02 „	
56	0·03 „	
78	0·04 „	
100	0·05 „	This load was raised from the bar several times, but permanent set was not sensible.
120	0·06 „	Every time the bar was relieved of this load, a set of about ·005 was observed.
200	0·17 „	
230	0·34 „	
320 . .	slipped through between the supports, bent nearly 3 inches, but not broken.	

We may therefore consider 100 lbs. as the utmost that the bar would support without permanent alteration, which is equivalent to a strain of 10,285 lbs. upon a square inch; and an extension of $\frac{1}{960}$th part of its length (see art. 110 and 121) Absolute cohesion greater than 34,000 lbs. for a square inch.

Calculating from this experiment, we find the weight of the modulus of elasticity for a base 1 inch square, 9,873,000 lbs.; and the specific gravity of gun-metal is 8·152; therefore the height of the modulus in feet is 2,790,000 feet.

The deflexion increases much more rapidly than in proportion to the weight, as soon as the strain exceeds the elastic force; a weight of 200 lbs. more than trebled the deflexion produced by 100, instead of only doubling it.

EXPERIMENTS ON BRASS.

97. Dr. Young made some experiments on brass, from which he calculated the height of the modulus of elasticity of brass plate to be 4,940,000 feet, or 18,000,000 lbs. for its weight to a base of 1 square inch. For wire of inferior brass he found the height to be 4,700,000 feet.[23]

As cast brass had not been submitted to experiment, I procured a cast bar of good brass, and made the following experiment:

The bar was filed true and regular; its depth was 0·45 inch, and breadth 0·7 inch. The distance between the supports 12 inches, and the scale suspended from the middle.

12 lbs. bent the bar	0·01 inch.	
23	0·02 ,,	
38	0·03 ,,	The bar was relieved several times, but it took no perceptible set.
52	0·04 ,,	
65	0·05 ,,	relieved, the set was ·01.
110	0·18 ,,	
163 . .	slipped between the supports, bent more than 2 inches, but not broke.	

Hence 52 lbs. seems to be about the limit which could not be much exceeded without permanent change of structure. It is equivalent to a strain of

[23] Natural Philosophy, vol. ii. p. 86.

6700 lbs. upon a square inch, and the corresponding extension is $\frac{1}{1333}$ of its length, (see art. 110 and 121). Absolute cohesion greater than 21,000 lbs. per square inch. The modulus of elasticity according to this experiment is 8,930,000 lbs. for a base of an inch square. The specific gravity of the brass is 8·37, whence we have 2,460,000 feet for the height of the modulus.

SECTION VII.

OF THE STRENGTH AND DEFLEXION OF CAST IRON WHEN IT RESISTS PRESSURE OR WEIGHT.

98. The doctrine of the Strength of Materials, as given in this Work, rests upon three first principles, and these are abundantly proved by experience.

The First is, that the strength of a bar or rod to resist a given strain, when drawn in the direction of its length, is directly proportional to the area of its cross section; while its elastic power remains perfect, and the direction of the force coincides with the axis.

99. The Second is, that the extension of a bar or rod by a force acting in the direction of its length, is directly proportional to the straining force, when the area of the section is the same; while the strain does not exceed the elastic power.[1]

[1] This limit should be carefully attended to, for as soon as the strain exceeds the elastic power, the ductility of the material becomes sensible. The degrees of ductility are extremely variable

100. The Third is, that while the force is within the elastic power of the material, bodies resist extension and compression with equal forces.

101. It is further supposed that every part of the same piece of the material is of the same quality, and that there are no defects in it. If there be any material defect in a piece of cast iron, it may often be discovered, either by inspection, or by the sound the piece emits when struck; except it be air-bubbles, which cannot be known by these means.

The manner of examining the quality of a piece of cast iron has been given in the Introduction, p. 7; and such as will bear the test of hammering with the same apparent degree of malleability, will be found sufficiently near of the same strength and extensibility for any practical deductions to be correct.

The truth of these premises being admitted, every rule that is herein grounded upon them may be considered as firmly established as the properties of geometrical figures.

102. A free weight or mass of matter is always to be considered to act in the direction of a vertical line passing through its centre of gravity; and its whole effect as if collected at the point where this

---

in different bodies, and even in different states of the same body. A fluid possesses this property in the greatest degree, for every change in the relative position of its parts is permanent.

vertical line intersects the beam or the pillar, which is to support it. But if the weight or mass of matter be partially sustained, independently of the beam or pillar, in any manner, then the direction and intensity of the force must be found that would sustain the mass in equilibrium,[2] and this will be the direction and intensity of the pressure on the beam or pillar.

103. Let f denote the weight in pounds which would be borne by a rod of iron, or other matter, of an inch square, when the strain is as great as it will bear without destroying a part of its elastic force.[3] Also, let W be any other weight to be supported, and $b =$ the breadth and $t =$ the thickness of the piece to support it, in inches. Then, by our first principles, art. 98, we have

$$f : W :: 1 : b\,t$$

$$\text{or, } \frac{W}{f} = b\,t \qquad \text{(i.)}$$

That is, the area should be directly as weight to be supported, and inversely as the force which would impair the elastic power of the material.

---

[2] The method of finding this force and its direction is explained in my Elementary Principles of Carpentry, art. 24-29.

[3] " A permanent alteration of form," Dr. Young has remarked, " limits the strength of materials with regard to practical purposes, almost as much as fracture, since in general the force which is capable of producing this effect is sufficient, with a small addition, to increase it till fracture takes place." Natural Philosophy, vol. i. p. 141.

104. If ϵ be the quantity a bar of iron, or other matter, an inch square, and a foot in length, would be extended by the force f; and l be any other length in feet; then

$$1 : l :: \epsilon : \Delta, \text{ or } l\epsilon = \Delta =$$

the extension in the length l. (ii.)

For, when the force is the same, the extension is obviously proportional to the length.

And, since by our principle, art. 99, the extension is as the force; we have $f : \mathrm{W} :: \epsilon$: extension produced by the weight $\mathrm{W} = \frac{\mathrm{W}\epsilon}{f}$, and we obtain from Equation ii. $\frac{\mathrm{W}l\epsilon}{f} = \Delta$. (iii.)

In which Δ is the extension that would be produced in the length l, by the weight W.

105. Where a comparison of elastic forces is to be made, it is sometimes convenient to have a single measure which is called the modulus of elasticity.[4] It is found by this analogy: as the length of a substance is to the diminution of its length, so is the modulus of elasticity to the force producing that diminution. Or, denoting the weight of the modulus in ℔s. for a base of an inch square by m,

$$\epsilon : f :: 1 : m = \frac{f}{\epsilon}. \qquad \text{(iv.)}$$

And if p be the weight of a bar of the substance

---

[4] The term was first used by Dr. Young. Lectures on Nat. Phil. vol. ii. art. 319.

1 foot in length, and 1 inch square; then if M be the height of the modulus of elasticity in feet,[5]

$$\frac{f}{p\,\epsilon} = \text{M}. \tag{v.}$$

106. Let the rectangular beam A A′, fig. 14, be supported upon a fulcrum D, in equilibrio, and for the present considering the beam to be acted upon by no other forces than the weights W, W′; which are supposed to have produced their full effect in deflecting the beam, and the vertical section at B D to be divided into equal, and very thin filaments, as shown in fig. 15.

Consider B, fig. 14, to be the situation of one of the small filaments in the upper part of the beam, and $a\,a'$ a tangent to the curvature of the filament B, at the point B. Now, it is clearly a necessary consequence of equilibrium that the forces tending to separate the filament at B should be equal, and in the direction of the tangent $a\,a'$; and the strain is obviously a tensile one.

But since F A is the direction of the weight, we have, by the principles of statics,

$\text{B}a : \text{A}a :: \text{S}$ (=the resistance of the filament B) : $\frac{\text{A}a.\text{S}}{\text{B}a}$ = its effect in sustaining the weight W.

These forces, we know both from reasoning and experience, will compress the lower part of the

[5] By this and the preceding equation, were calculated the height and weight of the modulus of elasticity of the different bodies in the Alphabetical Table.

beam; and let D be a compressed filament, of the same area as the filament B, and in the same position, and at the same distance from the under surface as the filament B is in respect to the upper surface. Also, let $e\,e'$ be a tangent to the filament at D, and parallel to $a\,a'$; and representing one of the equal and opposite strains on the filament D by e D; we have, $e\,\mathrm{D} : e\,\mathrm{A} :: \mathrm{S}'$ (the resistance to compression of D) : $\frac{e\,\mathrm{A}.\,\mathrm{S}'}{e\,\mathrm{D}}$ = the effect of the filament D in sustaining the weight W.

The effect of both the filaments, B D, in supporting the weight will therefore be,

$$\frac{\mathrm{A}\,a.\,\mathrm{S}}{\mathrm{B}\,a} + \frac{e\,\mathrm{A}.\,\mathrm{S}'}{e\,\mathrm{D}},$$

or since $\mathrm{B}\,a = e\,\mathrm{D}$, and as portions of the same matter of equal area resist extension or compression with equal forces (art. 100) $\mathrm{S} = \mathrm{S}'$; therefore, $\frac{\mathrm{S}}{\mathrm{B}\,a} \times (\mathrm{A}\,a + e\,\mathrm{A})$ = the effect of the filaments D and B.[6] But

$$\mathrm{A}\,a + e\,\mathrm{A} = \mathrm{B}\,\mathrm{D},$$

[6] But when the strain exceeds the elastic force of a body, the resistance to compression exceeds the resistance to tension; consequently, the effect of the filaments must be

$$\frac{\mathrm{A}\,a.\,\mathrm{S} + e\,\mathrm{A}.\,\mathrm{S}'}{\mathrm{B}\,a}.$$

Now the difference between S and S′ will be constantly increasing till fracture takes place, the area of the compressed part being constantly increasing, and that of the extended part diminishing. The variation will depend on the ductility of the material, but it

the vertical distance between the filaments.[7] Consequently

$$\frac{S.BD}{Ba} = \text{this effect in supporting the weight W.} \quad \text{(vi.)}$$

107. As one side of the beam suffers extension, and the other side compression, there will be a filament at some point of the depth, which will neither be extended nor compressed; the situation of this filament may be called the neutral axis, or axis of motion.

The extension or compression of a filament will obviously be as its distance from the neutral axis; and when the neutral axis divides the section into two equal and similar parts, its place will be at the middle of the depth.

And since the effect of two equal filaments is as the distance between them, the effect of either will be as its distance from the neutral axis; for the filaments being equal, and the strain on them equal, the axis will be at the middle of the distance between them; and the effect of both being measured

---

cannot be ascertained, unless some very careful experiments were made for the purpose; and fortunately it is an inquiry not required in the practical application of theory.

[7] When the flexure becomes considerable, the curve is flattened in consequence of the forces compressing the beam, and A a + e A will exceed the vertical distance between the filaments; and the point of greatest strain will be found to change to the place where the line A B intersects the filament. This change of the point of greatest strain is very apparent in experiment.

by the whole depth, that of one of them will be measured by half the depth. Therefore the effect of a filament is

$$\frac{S.BD}{2\,(B\,a)} = \frac{S.B\,d}{B\,a}. \qquad \text{(vii.)}$$

108. When a beam is sustained in any position,[8] not greatly differing from a horizontal one, by a fulcrum, as in fig. 14, the power of a fibre or filament to support a weight at A or A′ is directly as its force, its area, and the square of its distance from the neutral axis; and inversely as the distance, F B, of the straining force from the point of support.

For the strain being as the extension, and the extension of any filament being directly as the distance of that filament from the axis of motion, therefore, the force of a filament is as its distance from the axis of motion. But it has been shown (art. 103,) that the force is also as the area; and the power in sustaining a weight has been shown (art. 107) to be directly as the vertical distance from the neutral axis, and inversely as the length B a, that is, as $\frac{B\,d}{B\,a}$; and, since the triangles F B a, B df, are similar,

$$\frac{B\,d}{B\,a} = \frac{f\,d}{F\,B}, \text{ therefore}$$

---

[8] It does not sensibly differ from the correct law of resistance till the beam be so much inclined as to slide on its support; but the general investigation will be found in art. 276.

$\frac{(f\,d)^2 \times \text{the force of filament} \times \text{by its area}}{\text{F B}} = $ the weight it will sustain. (viii.)

109. Let d be the depth, divided into filaments, each equal to x the mth part of $\frac{d}{2}$; also put F B $= l$, the breadth of the beam $= b$, and f the weight that a fibre of a given magnitude would bear when drawn in the direction of its length, without destroying its elastic force.

Now, if we calculate the mean strain upon each filament by Equation viii. art. 108, we obtain the following progression, and its sum is the weight the beam will support.[9]

$$\frac{4\,f\,b\,x^3}{l\,d} \times (1 + 2^2 + 3^2 \ldots\ldots \overline{m-1}^2 + \frac{m^2}{2}) = \text{W}. \qquad \text{(ix.)}$$

110. If the beam be rectangular, the value of

$$\text{W} = \frac{f\,b\,d^2}{6\,l}. \qquad \text{(x.)}$$

Therefore the lateral strength of a rectangular beam is directly as its breadth, and the square of its depth, and inversely as its length.

And when the beam is square, its lateral strength is as the cube of its side.

111. If a plate be fixed along one of its sides A B, and the load be applied at the angle C; then,

---

[9] The first term of the progression is equivalent to the quantity called a fluxion, and is usually written thus, $\frac{4\,f\,b\,x^2\,\dot{x}}{l\,d}$. The same remark applies to the other progressions.

if the distance A B be greater than B C, the plate will break in the direction of some line E B. To find this line, put F C = l, the leverage, and E B = b, the breadth; also t = the tangent of the angle E B C. Then, by similar triangles,

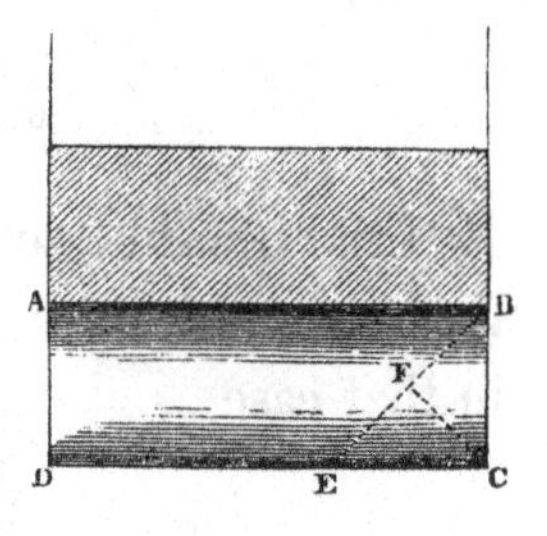

$$\sqrt{1 + t^2} :: \mathrm{BC} : l = \frac{\mathrm{BC} \times t}{\sqrt{1 + t^2}};$$

and

$$1 : \sqrt{1 + t^2} :: \mathrm{BC} : b = \mathrm{BC} \times \sqrt{1 + t^2};$$

therefore

$$\frac{f\,b\,d^2}{6\,l} = \frac{f\,d^2\,(1 + t^2)}{6\,t}.$$

But this equation is a minimum when $t = 1$; that is, when the angle E B C is 45 degrees; consequently

$$\frac{f\,d^2\,(1 + t^2)}{6\,t} = \frac{f\,d^2}{3} = \mathrm{W}. \qquad \text{(xi.)}$$

112. If the beam be rectangular, and the strain be in a perpendicular direction to one of its diagonals A C, making that diagonal = b, and the depth E F = a, the progression becomes (because the breadth is successively

$\overline{a - 2\,x}$, $\overline{a - 4\,x}$, &c.)

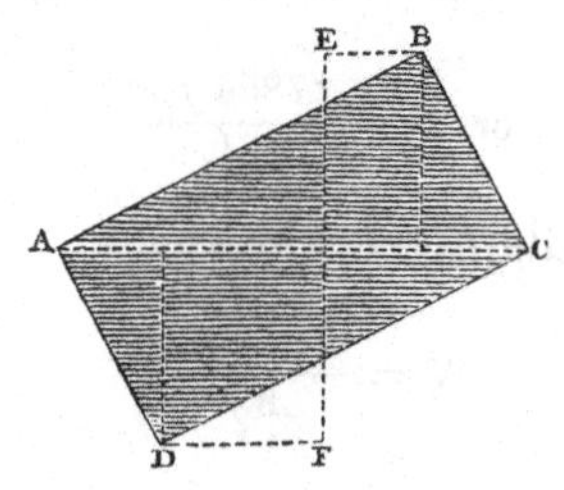

$$\frac{4\,b\,f\,x^3}{l\,a} \times \left\{ a\,(1+2^2+\ldots\ldots\frac{m^2}{2}) - 2\,x\,(1+2^3+\ldots\frac{m^3}{2}). \right\}$$

$$\text{or, } W = \frac{f\,b\,a^2}{24\,l}. \qquad \text{(xii.)}$$

If the beam be square, the direction of the straining force coincides with the vertical diagonal, and in that case

$$\frac{f\,a^3}{24\,l} = W. \qquad \text{(xiii.)}$$

But the diagonal of a square beam is equal to its side multiplied by $\sqrt{2}$; hence, if d be the side, we have

$$\frac{f\,d^3}{6\,\sqrt{2}\,l} = W. \qquad \text{(xiv.)}$$

Consequently the strength of a square beam, when the force is parallel to its side, is to the strength of the same beam when the force is in the direction of its diagonal as

$$1 : \frac{1}{\sqrt{2}};$$

or as 10 is to 7 nearly.

113. If the beam be a cylinder, and r the radius, then b is successively

$$2\,\sqrt{r^2 - x^2},\ 2\,\sqrt{r^2 - (2\,x)^2},\ \&c.,$$

And $\frac{4\,f\,x^3}{r\,l} \times \left\{ \sqrt{r^2 - x^2} + 2^2\,\sqrt{r^2 - (2\,x)^2} + \&c. \right\} = W;$

$$\text{or, } W = \frac{\cdot 7854\,f\,r^3}{l}. \qquad \text{(xv.)}$$

If d be the diameter, then

$$W = \frac{\cdot 7854\,f\,d^3}{8\,l}. \qquad \text{(xvi.)}$$

The lateral strength of a cylinder is directly as the cube of its diameter, and inversely as the length.

The strength of a square beam is to that of an inscribed cylinder as

$$8 : 6 \times \cdot 7854, \text{ that is, as } 1 : \cdot 589, \text{ or as } 1 \cdot 7 : 1.$$

114. If the section of the beam be an ellipse, when the strain is in the direction of the conjugate axis, we have by the same process

$$W = \frac{\cdot 7854 f t c^2}{l}, \qquad \text{(xvii.)}$$

where t is the semi-transverse, and c the semi-conjugate axis.

115. If the beam be a hollow cylinder or tube, and r be the exterior radius, $n r$ being that of the hollow part, then by the same process [10] we find

$$W = \frac{\cdot 7854 f r^3 (1 - n^4)}{l}. \qquad \text{(xviii.)}$$

The radius of a solid cylinder that will contain the same quantity of matter as the tube, is easily found by geometrical construction in this manner: make B D perpendicular to B C; then C D being the radius of the tube, and B C that of the hollow part, B D will be the radius of a solid cylinder

[10] Dr. Young gives a rule which is essentially the same, of which I was not aware when my 'Principles of Carpentry' was written. See Natural Philosophy, vol. ii. art. 339, B. scholium. In a recent work on the Elements of Natural Philosophy, by Professor Leslie, vol. i. p. 242, the learned author has neglected to consider the effect of extension in his investigation of this equation.

which will contain the same quantity of matter as the tube. Because

$$BD^2 = CD^2 - BC^2.$$

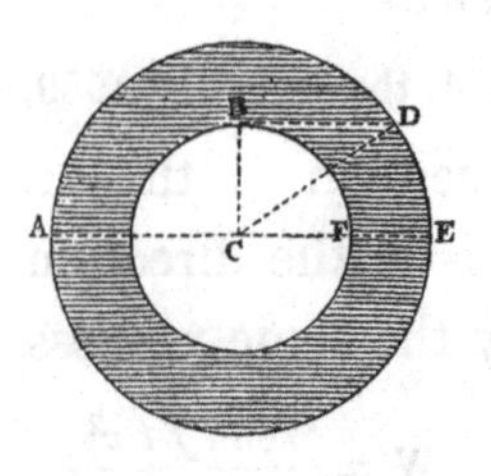

By comparing Equation xv. and xviii. we find that when a solid cylinder is expanded into a tube, retaining the same quantity of matter, the strength of the solid cylinder being 1, that of the tube will be $\frac{1 - n^4}{(1-n^2)^{\frac{3}{2}}}$. When the thickness FE is $\frac{1}{5}$th of the diameter AE, the strength will be increased in the proportion of 1·7 to 1. And when FE is $\frac{3}{20}$ths of the diameter, the strength will be doubled by expanding the matter into a tube. But a greater excess of strength cannot be safely obtained than the latter, because the tube would not be capable of retaining its circular form with a less thickness of matter. From $\frac{1}{6}$th to $\frac{1}{10}$th seems to be the most common ratio in natural bodies, such as the stems of plants, &c.

116. If a beam be of the form shown in Plate I. fig. 9, (see art. 38, 39, and 40,) and d be the extreme depth, and b the extreme breadth; $q\,b =$ the difference between the breadth in the middle and the extreme breadth, and $p\,d$ the depth of the narrow

part in the middle; then by the process employed in calculating Equation x. we find

$$W = \frac{f\,b\,d^2}{6\,l} \times (1 - q\,p^3.) \qquad \text{(xix.)}$$

117. If the middle part of the beam be entirely left out, with the exception of cross parts to prevent the upper and lower sides coming together, as in figs. 11 and 12, Plate II. (see art. 41,) and d be the whole depth, $p\,d$ the depth of the part left out in the middle, and b the breadth, then

$$W = \frac{f\,b\,d^2}{6\,l}(1 - p^3.) \qquad \text{(xx.)}$$

118. Hitherto we have only considered those forms where the neutral axis divides the section into identical figures; but there are some interesting cases [11] where this does not happen, such, for example, as the triangular section.

Taking the case of a triangular section with a part removed at the vertex, we shall have a general case which will include that of the entire triangle. Let d be the depth of the complete triangle, and $m\,d$ the depth of the part cut off at the vertex; and $n\,d$ the depth of the neutral axis, M N, from the upper side $a\,b$ of the beam; then the distance of the neutral axis from the base will be $(1-n-m)\,d$. If both sides of the neutral axis were the same as

[11] They are interesting, because the early theorists fell into some serious errors respecting them, and consequently have led practical engineers into erroneous opinions.

the upper one, the strength would be equal to that

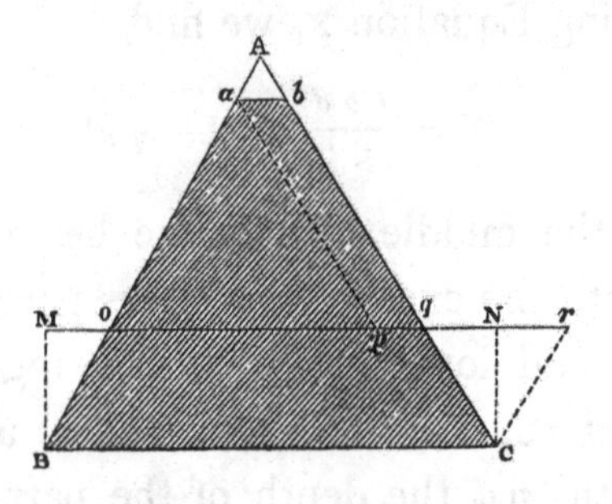

of a parallelogram $a\,b\,p\,q$, added to a triangle $a\,o\,p$; hence from Equation x. and xii. we have

$$\frac{f}{6\,l}(4\,m\,b\,n^2\,d^2 + b\,n^3\,d^2) = \frac{f\,b\,n^2\,d^2}{6\,l}(4\,m + n) = \mathrm{W}.$$

But to find the place of the neutral axis, we must compare the strength of the lower side with the upper one; and the strength of the lower side is equal to that of a rectangle M N B C minus a triangle $q\,r$ C, or

$$\frac{f}{6\,l}\left\{4\,b\,d^2\,(1 - m - n)^2 - b\,d^2\,(1 - m - n)^3\right\} = \mathrm{W}.$$

And consequently,

$$n^2\,(4\,m + n) = 4\,(1 - m - n)^2 - (1 - m - n)^3.$$

Whence,

$$n = \frac{5-2\,m-3\,m^2}{2\,(1-m)} - \sqrt{\left(\frac{5-2\,m-3\,m^2}{2\,(1-m)}\right)^2 - \frac{3-5\,m+m^2+m^3}{1-m}}.$$

When $m = 0{\cdot}1$, then $n = {\cdot}592$, and

$$\frac{{\cdot}348\,f\,b\,d^2}{6\,l} = \mathrm{W}. \qquad \text{(xxi.)}$$

But if $m = o$, or the triangle A B C be entire, then $n = \cdot 697$ nearly.[12] And

$$\frac{\cdot 339 f b d^2}{6 l} = W, \text{ or } \frac{\cdot 0565 f b d^2}{l} = W.^{13} \qquad \text{(xxii.)}$$

If $m = \frac{1}{9}$ we have $n = \cdot 58166$, and

$$\frac{\cdot 347 f b d^2}{6 l} = W. \qquad \text{(xxiii.}$$

Where $m = 0\cdot 1$, the strength is about the greatest possible, a triangular prism being about $\frac{1}{37}$th part stronger when the angle is taken off to $\frac{1}{10}$th of its depth, as shown by the shaded part of the figure. Emerson first announced this seeming paradox,[14] but it is easily shown that his solution only applies to the imaginary case where the neutral axis is an incompressible arris, at the base of the section.

A triangular prism is equally strong, whether the base or the vertex of the section be compressed:[15] and by comparing Equations x. and xxii. it appears that its strength is to that of a circumscribed rectangular prism as 339 : 1000, or nearly as 1 : 3. But let it be remembered that this ratio only applies to strains which do not produce permanent

---

[12] Duleau obtains a result equivalent to $n = \cdot 57$ in this case, but the result only is given. Essai Théorique, &c. p. 77.

[13] This rule was first published in the Philosophical Magazine, vol. xlvii. p. 22. 1816.

[14] Mechanics, Sect. VIII. p. 114. 5th edit. 1800.

[15] Duleau has proved the truth of this by his experiments on the flexure of triangular bars. Essai sur la Résistance, &c. p. 26.

alteration in the materials, and where the arris is not injured by the action of the straining force: if the strain be increased so as to produce fracture, the triangle will be found still weaker than in this proportion, when the arris is extended, and somewhat stronger when the arris is compressed; in the former case, from the imperfection of castings, where there is much surface in proportion to the quantity of matter, as in all acute arrises; in the latter case, from the saddle or other thing used to support the weight reducing the quantity of actual leverage.

It may be useful to remark, that a triangle contains half the quantity of matter that there is in the circumscribing rectangle, but its strength is only one-third; hence it is not economical to adopt triangular sections, and a like remark applies to the T formed sections so commonly used.

119. If the whole depth of a T formed section be d, its greatest breadth b, and its least breadth $(1-q)\,b$. Then, supposing the depth of the neutral axis M N from the narrow edge A E to be $\frac{1}{n}\,d$, the strength of the bar will be

$$\frac{4\,f\,b\,d^2\,(1-q)}{6\,l\,n^2} = \mathrm{W}.$$

For $\frac{4\,d^2}{n^2}$ would be the square of the whole depth were both sides of the neutral axis the same; and the strength would be equal to a rectangle of that depth with the breadth $(1-q)\,b$.

But the strength of the other side of the axis by Equation xix. is

$$\frac{4 f b d^2 (1 - q p^3)(n - 1)^2}{6 l n^2};$$

hence we have the equation $(n - 1)^2 (1 - q p^3) = 1 - q$ to determine the place of the neutral axis; or

$$n = 1 + \sqrt{\frac{1 - q}{1 - q p^3}};$$

consequently,

$$\frac{4 f b d^2 (1 - q)}{6 l \left(1 + \sqrt{\frac{1 - q}{1 - p^3 q}}\right)^2} = \mathrm{W}. \qquad \text{(xxiv.)}$$

This formula is complicated, but it affords some curious results. If we make $p = o$, we have the strength of a bar with its neutral axis at C, and the depth A C is

$$d \left(\frac{1}{1 + \sqrt{1 - q}}\right);$$

where d is the whole depth.

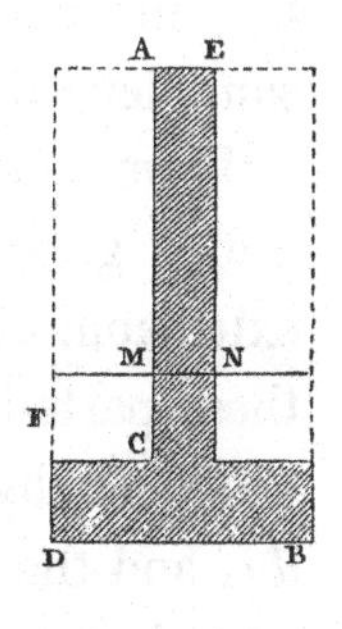

And if $\mathrm{A E} = \frac{1}{4}\,\mathrm{D B}$, then $\mathrm{A C} = \frac{2}{3}\,d$, when the neutral axis is at C.

The neutral axis may be at any point that may be chosen between the point C and half the depth, by varying the values of q or p for that purpose.

If we make

$$q = \cdot 75, \text{ and } p = \cdot 5, \text{ then } \mathrm{A M} = \frac{d}{1 \cdot 55}, \text{ and } \mathrm{D F} = \mathrm{C M};$$

also,

$$A\,E = \tfrac{1}{4}\,D\,B.$$

The strength is

$$\frac{f\,b\,d^2}{6 \times 2{\cdot}4\,l} = W. \qquad \text{(xxv.)}$$

The figure is drawn in these proportions, b is the whole breadth D B, and d the whole depth. Its strength is to that of the circumscribing rectangle shown by the dotted lines as $\frac{1}{2{\cdot}4} : 1$ or as $5 : 12$.

These equations show the relation between the strength of beams, and the weight to be supported in some of the most useful cases when the load is applied as in fig. 14; but previous to considering how these equations will be effected by varying the mode of supporting the beam, it will be desirable to give some rules for estimating the deflexion of beams.

120. The deflexion of a beam supported as in fig. 16, Plate II., is caused by the extension of the fibres of the upper side, and the compression of those on the under side; the neutral line A B A′ retains the same length.

If we conceive the length of a beam to be divided into a great number of equal parts, and that the extension, at the upper side of the beam, of one of these parts is represented by $a\,b$, then the deflexion produced by this extension will be represented by $d\,e$, and the angles $a\,c\,b$, $d\,c\,e$, being equal, we shall have $b\,c : d\,c :: a\,b : d\,e$; the smallness of the angles rendering the deviation from strict similarity insensible.

Now, however small we may consider the parts to be, into which the length is divided, still the strain will vary in different parts of it, and consequently the deflexion; but if we consider the deflexion produced by the extension of any part, to be that which is due to an arithmetical mean between the greatest and least strains in that part, we shall then be extremely near the truth.

We have seen that the strain is as the weight and leverage directly, and as the breadth and square of the depth inversely, (see art. 108). Our investigation will be more general by considering the weight, breadth, and depth variable, by taking l, b, and d for the length, breadth, and depth of the middle or supported point, and W for the whole weight, and x, y, and w for the depth, breadth, and weight on any other point. Then, the deflexion from the strain at any point c is as

$$\frac{W\,l\,x}{2\,b\,d^2} : \frac{w \times (d\,c)^2}{y\,x^2} :: \epsilon : \frac{2\,b\,d^2\,\epsilon\,w\,(d\,c)^2}{W\,l\,y\,x^3}.$$

And, if z be the length of one of the parts into which we suppose the whole length divided, then the deflexion from the mean force on the length of z situate at c will be

$$\frac{2\,b\,d^2\,\epsilon\,w\,z}{W\,l\,y\,x^3} \times \left(\frac{d\,c^2 + \overline{d\,c + z}^2}{2}.\right)$$

Since the whole deflexion D A is the sum of the deflexions of the parts, we have

$$\frac{2\,b\,d^2\,\epsilon\,w\,z^3}{W\,l\,y\,x^3} \times \left(1^2 + 2^2 + \&c.\ \overline{m-1}^2 + \frac{m^2}{2}\right) = \text{D A}. \qquad \text{(i.)}$$

121. *Case* 1. When a beam is rectangular, the depth and breadth uniform, and the load applied at one end. Then,

$$b = y,\ d = x, \text{ and } W = w.$$

Therefore the progression becomes

$$\frac{2 \epsilon z^3}{l d} \times (1^2 + 2^2, \&c.),$$

of which the sum is

$$\frac{2 \epsilon l^2}{3 d} = \text{the deflexion D A.}^{16} \qquad \text{(ii.)}$$

122. *Case* 2. When the section of the beam is rectangular, and the load acts at one end, the depth being uniform, but the breadth varying as the length.

In this case the progression is

$$\frac{2 \epsilon z^2}{d} \times (1 + 2, \&c.) = \frac{\epsilon l^2}{d} = \text{the deflexion D A.} \qquad \text{(iii.)}$$

This is the beam of uniform strength described in art. 30, fig. 6, and the deflexion is $\frac{1}{3}$rd more than that of a beam of equal breadth throughout its length. The deflexion of the beam of equal strength described in art. 33 is the same; the neutral axis becomes a circle in both these beams.[17]

---

[16] The same relation is otherwise determined in Dr. Young's Natural Philosophy, vol. ii. art. 325.

[17] M. Girard arrives at the erroneous conclusion, that all the solids of equal resistance curve into circular arcs, (Traité Analytique, p. 82,) in consequence of neglecting the effect of the depth of the solid on the radius of curvature.

In this case it is easily shown by other reasoning that the curve of the neutral axis is a portion of a circle, and it is well known that in an arc of very small curvature, (one of such as are formed by the deflexions of beams in practical cases,) the versed sine is sensibly proportioned to the square of the sine. This will enable the reader to form an estimate of the accuracy of the method I here follow. I am perfectly satisfied that it is correct enough for use in the construction of machines or buildings, and that it is an useless refinement to embarrass the subject with intricate rules; but this explanation may be necessary to some nice theorists, who aim rather at imaginary perfection than useful application.

123. *Case* 3. When the section of the beam is rectangular, the load acting at the end, the breadth uniform, and the depth varying as the square root of the length; which is the parabolic beam of equal strength. (See art. 27, fig. 3, Plate I.)

In this case the progression is

$$\frac{2\,\epsilon\, l^{\frac{1}{2}}\, z^{\frac{3}{2}}}{d} \times \left(1^{\frac{1}{2}} + 2^{\frac{1}{2}} + \text{\&c.}\right) \text{or,}$$

$$\frac{4\,\epsilon\, l^2}{3\,d} = \text{the deflexion D A.} \qquad \text{(iv.)}$$

The deflexion is double that of an uniform beam while the quantity of matter is only lessened $\frac{1}{3}$rd.

124. *Case* 4. When the section of the beam decreases from the supported point to the end where

the load acts, so that the sections are similar figures, then the curve bounding the sides of the beam will be a cubical parabola; that is, the depth will be every where proportional to the cube root of the length.

In this case the progression is

$$\frac{2\,\epsilon\,l^{\frac{1}{3}}z^{\frac{5}{3}}}{d} \times (1^{\frac{2}{3}} + 2^{\frac{2}{3}} + \&c.) = \frac{6\,\epsilon\,l^2}{5\,d} = \text{the deflexion D A.} \quad \text{(v.)}$$

The deflexion is to that of an uniform beam as 1·8 : 1.

125. *Case* 5. When a beam is of the same breadth throughout, and the vertical section is an ellipse, (see fig. 8, art. 32,) the deflexion from a weight at the vertex may be exhibited in a progression as below:

$$\frac{l^2\,\epsilon\,z^3}{d} \times \left\{\frac{2}{(2\,l\,z - z^2)^{\frac{3}{2}}} + \frac{8}{(4\,l\,z + 4\,z^2)^{\frac{3}{2}}} + \&c. + \frac{m^2}{(2lmz - m^2z^2)^{\frac{3}{2}}}\right\}$$

$$= \frac{l^2\,\epsilon\,z^{\frac{3}{2}}}{d} \times \left(\frac{2}{(2\,l - z)^{\frac{3}{2}}} + \&c.\right) = \text{D A.}$$

By actually summing this progression when $m = 10$, we have

$$\frac{\cdot 857\,l^2\,\epsilon}{d} = \text{the deflexion D A.} \qquad \text{(vi.)}$$

126. *Case* 6. If a rectangular beam of uniform breadth and depth be so loaded that the strain be upon any point c, then

$$l^2 : d\,c \times (2\,l - d\,c) :: \text{W} : w = \frac{d\,c \times \text{W} \times (2\,l - d\,c)}{l^2}.$$

This value of w being substituted in Equation i., we have

$$\frac{\epsilon z^3}{d l^2}\left\{2 l (1^2 + 2^3 + \&c.) - z (1^3 + 2^3, \&c.)\right\} = \frac{\epsilon l^2}{d} \times$$

$$(\tfrac{4}{3} - \tfrac{1}{2}) = \frac{5 \epsilon l^2}{6 d} = \text{the deflexion D A.}^{18} \qquad \text{(vii.)}$$

This is the deflexion of a beam uniformly loaded when it is supported at the ends, l being half the length.

127. *Case* 7. If the section of a beam be rectangular, and the breadth uniform, but a portion of the depth varying as the length, and the rest of it uniform: then the depth at any point c will be

$$\frac{d}{l}(\overline{1 - n}\, l + n z) = x;$$

the depth at the point where the weight acts being the $\overline{1-n}$th part of the depth at the point of support.

This value of x being substituted in Equation i. art. 120, it becomes

$$\frac{2 l^2 \epsilon z^3}{d} \times \left\{\frac{1}{(\overline{1-n}\, l + n z)^3} + \frac{2^2}{(\overline{1-n}\, l + 2 n z)^3} + \&c.\right\} = \text{D A.}$$

And the general expression for the sum of this progression is

$$\frac{2 l^2 \epsilon}{d}\left\{\frac{2(1-n)}{n^2} + \frac{3(1-n)^2}{2 n^3} - \frac{3}{2 n^3} + \frac{1}{n^3}\text{hy. log.}\frac{1}{1-n}\right\} = \text{D A.}$$

When $n = \cdot 5$, as in the beams figs. 4 and 5, Plate I., we have

---

[18] This relation is otherwise determined by Dr. Young, Nat. Philos. vol. i. art. 329.

$$\frac{1{\cdot}09\, l^2\, \epsilon}{d} = \text{D A the deflexion.} \qquad \text{(viii.)}$$

Hence the deflexion of an uniform beam being denoted by 1, this beam will be deflected 1·635 by the same force; the middle sections being the same.

If the beam be diminished at the end to two-thirds of the depth at the middle, then

$$n = \frac{1}{3};$$

and

$$\frac{0{\cdot}895\, l^2\, \epsilon}{d} = \text{D A the deflexion.} \qquad \text{(ix.)}$$

128. *Case* 8. If a rectangular beam be supported in the middle and uniformly loaded over its length; then,

$$l : (d\, c) :: \text{W} : w = \frac{\text{W}\, (d\, c)}{l}.$$

Hence, when the beam is of uniform breadth and depth, we have by substituting this value of w in Equation i. art. 120,

$$\frac{2\, \epsilon\, z^4}{l^2} \times (1 + 2^3 + 3^3 + \&\text{c.}) = \frac{l^2\, \epsilon}{2\, d} = \text{D A the deflexion.} \qquad \text{(x.)}$$

In this case the deflexion is $\frac{3}{4}$ths of the deflexion of the same beam having the whole weight collected at the extremities.

129. *Case* 9. If a beam be generated by the revolution of a semi-cubical parabola round its axis, which is the figure of equal strength for a beam supported in the middle when the weight is uniformly diffused over its length. Then,

$$l^2 : d^3 :: (d\, c)^2 : x^3 = \frac{d^3\, (d\, c)^2}{l^2};$$

$$\text{and } y = x; \text{ also } w = \frac{(d\,c)\,W}{l}.$$

These quantities, substituted in Equation i. art. 120, give

$$\frac{2\,\epsilon\, l^{\frac{2}{3}} \cdot z^{\frac{4}{3}}}{d} \times \left\{1 + 2^{\frac{1}{3}} + 3^{\frac{1}{3}} + \&c.\right\} = \frac{3\,\epsilon\, l^2}{2\,d} = \text{D A}$$

the deflexion. (xi.)

Here the deflexion is $\frac{9}{4}$ths of that of an uniform beam with the load at the extremities.

130. *Case* 10. If a beam to support an uniformly distributed load be of equable breadth, but the depth varying directly as the distance from the extremity, as in fig. 21, Plate III. Then

$$l : d :: (d\,c) : x = \frac{d\,(d\,c)}{l};$$

$$b \text{ is constant, and } w = \frac{(d\,c)\,W}{l};$$

therefore by Equation i. art. 120,

$$\frac{2\,\epsilon\, l^2}{d} = \text{D A the deflexion.} \qquad \text{(xii.)}$$

If the beam had been uniform, and the loads at the extremities, the deflexion would have been only $\frac{1}{3}$rd of the deflexion in this case.

The cases I have considered are perhaps sufficient for the ordinary purposes of business; the next object is to show how these calculations are affected by changing the position and manner of supporting the beam, or the nature of the straining force; and to compare them with experiments, and draw them into practical rules. For this purpose the most clear and the most useful plan seems to be that of taking known practical cases for illustration.

BEAMS SUPPORTED IN THE MIDDLE, AND STRAINED AT THE ENDS, AS IN THE BEAM OF A STEAM ENGINE.

131. The distance F B, fig. 14, of the direction of the straining force from the centre of motion being constantly the same, the strain will be the same in any position of the beam (art. 108). Also, the deflexion from its natural form will be the same in every position, because the strain is the same; and the length does not vary with the position.

Now the force acting upon the beam of a steam engine being impulsive, the practical rules for its strength will be found in the eleventh section; the formula calculated in this section being used to establish those rules.

BEAMS FIXED AT ONE END; AS CANTILEVERS, CRANKS, &c.

132. The strain upon a beam supported upon a fulcrum, as in fig. 14, is obviously the same as when one of the ends is fixed in a wall, or other like manner; for fixing the end merely supplies the place of the weight otherwise required to balance the straining force. But though the strain upon the beam be the same, the deflexion of the point where the strain is applied will vary according to

the mode of fixing the end; because the deflexion of the strained point will be that produced by the curvature of both the parts A B and B A′.

133. Let the dotted lines in fig. 17, Plate III., represent the natural position of a beam fixed at one end in a wall: when this beam is strained by a load at A, the compression at C will always be enough to allow the beam to curve between A′ and B, and the strain at the point A′ will obviously be the same as if a weight were suspended there that would balance the weight at A. Let A B A′ be the curvature of the beam by the load W, and $a\,a'$ a tangent to the point B. Then A′ a' is proportional to the deflexion produced by the strain at A′, and

$$A'B : BD :: A'a' : Da = \frac{BD \times A'a'}{A'B} =$$

the deflexion from the curving of the part A′ B; therefore

$$\frac{BD \times A'a'}{A'B} + Aa = \text{the whole deflexion } DA.$$

Now, since the deflexion is as the square of the length, (see Equation i.-xii. art. 120-130,) we have

$$(BA)^2 : (BA')^2 :: Aa : A'a' = \frac{Aa \times (BA')^2}{(BA)^2}.$$

Therefore,

$$Aa \times \left(1 + \frac{BD \times BA'}{(BA)^2}\right) = DA. \qquad \text{(i.)}$$

If the angle D B A be represented by c, then

$$BD = BA \times \cos. c;$$

and putting

$$r = \frac{\text{B A}'}{\text{B A}},$$

we have

$$\text{A}\,a \times (1 + r.\ \cos.\ c) = \text{D A}. \qquad \text{(ii.)}$$

But since the deflexion is always very small, in practical cases, we may always consider cos. $c = 1$, or equal to the radius, and then we have

$$\text{A}\,a \times (1 + r) = \text{D A}. \qquad \text{(iii.)}$$

134. In this equation r is the ratio of the length out of the wall to the length within the wall; that is,

$$\text{B A} : \text{B A}' :: 1 : r.$$

If the beam be either supported in the middle on a fulcrum, or fixed so that the length of the fixed part be equal to that of the projecting part, then

$$r = 1, \text{ and } 2\,(\text{A}\,a) = \text{D A}. \qquad \text{(iv.)}$$

135. If the fixed part be of greater bulk than the projecting part, or it be so fixed that the extension of the fixed part would be very small, then the effect of such extension may be neglected, and the deflexion D A and A a will be the same; particularly in the cranks of machinery, as in fig. 18, because by employing this value of D A in calculating the resistance to impulsion, we err on the safe side. See art. 327.

BEAMS SUPPORTED AT BOTH ENDS, AS BEAMS FOR SUPPORTING WEIGHTS, &c.

136. When the same beam is supported at the ends, as in fig. 19, instead of being loaded at the

ends, and supported in the middle, as in fig. 14, and the inclination and sum of the load be the same in both positions, the strains will be the same.

In either position of the beam we have

$$W \times FB = W' \times F'B,$$

or as

$$W : W' :: F'B : FB;$$

and therefore,

$$W + W' : W :: FF' : F'B.$$[19]

Consequently,

$$W \times FB = \frac{\overline{W + W'} \times F'B \times FB}{FF'}.$$

If the beam be a rectangle, and the whole length $FF' = l$, and W the whole weight, then by art. 110, Equation x.

$$\frac{f b d^2}{6} = \frac{W \times FB \times F'B}{l}. \quad \text{(v.)}$$

137. And the strain is as the rectangle of the segments into which the point B divides the beam; and therefore the greatest when the point B is in the middle, as has been otherwise shown by writers on mechanics.[20]

If the weight be applied in the middle, then

$$\frac{\overline{W + W'} \times F'B \times FB}{FF'} = \frac{\overline{W + W'} \times FF'}{4}.$$

In a rectangular beam, the whole length being l, and W the whole weight, then

$$\frac{f b d^2}{6} = \frac{l W}{4}, \text{ or } \frac{2 f b d^2}{3} = l W. \quad \text{(vi.)}$$

---

[19] Euclid's Elements, Prop. XVIII. Book V.

[20] Gregory's Mechanics, vol. i. art. 178, cor. 2.

138. When a weight is distributed over the length of a beam A B, fig. 20, in any manner, the strain at any point C may be found. For let G be the centre of gravity of that part of the load upon A C, and g that of the load upon B C. Then by the property of the lever,

$\frac{w \times AG}{AC}$ = the stress at C from the weight w of the load upon A C.

Also,

$\frac{w' \times gB}{CB}$ = the stress at C from the weight w' of the load upon C B.

Therefore the whole stress is

$$\frac{\overline{w \times CB \times AG} + \overline{w' \times AC \times gB}}{AC \times CB}.$$

And by Equation v. art. 136, the strain will be

$$\frac{\overline{w \times CB \times AG} + \overline{w' \times AC \times gB}}{AB} \qquad \text{(vii.)}$$

139. *Case* 1. When the weight is uniformly distributed over the length, then

$$AG = \tfrac{1}{2} AC; \ gB = \tfrac{1}{2} CB, \text{ and } w + w' = W,$$

the whole weight upon the beam; these values being substituted in Equation vii. it becomes

$$\frac{W \times AC \times CB}{2\,AB} = \text{strain at C.} \qquad \text{(viii.)}$$

The strain is greatest at the middle of the length, for then A C×C B is a maximum, and it is evidently the same as if half the weight were collected there; for in that case A C being equal C B, and either of these equal to half A B, we have in the case of a rectangular beam

$$\frac{f\,b\,d^2}{6} = \frac{l\,W}{8}, \text{ or } \frac{4\,f\,b\,d^2}{3} = l\,W. \qquad \text{(ix.)}$$

140. *Case* 2. When the load increases from A to B in proportion to the distance from A; then

$$AG = \tfrac{2}{3}\,AC, \text{ and } g\,B = \tfrac{1}{3}\,CB \times \frac{3\,AB - 2\,CB}{2\,AB - CB}.$$

Now since

$$w + w' = \text{the whole weight,}$$

and

$$w = \frac{\tfrac{1}{2}\,AC^2 \times W}{AB},$$

also

$$w' = \tfrac{1}{2}\,CB \times W\,\frac{2\,AB - CB}{AB},$$

if these values be inserted in Equation vii. we have

$$\frac{W.AC}{6\,AB}\,(AB^2 - AC^2) = \text{the strain at C.} \qquad \text{(x.)}$$

By the principles of maxima and minima of quantities, we readily find that the strain is the greatest at the distance of $\sqrt{\tfrac{1}{3}\,AB}$ from A. And the strain will be nearly

$$\frac{AB^2.\,W}{15{\cdot}59} \text{ at the point of greatest strain} = \frac{W.\,AB}{7{\cdot}75} \text{ when W is the whole weight.} \qquad \text{(xi.)}$$

This distribution of pressure applies to the pressure of a fluid against a vertical sheet of iron; as in lock-gates, reservoirs, sluices, cisterns, piles for wharfs, &c.

141. *Case* 3. When the load increases as the

square of the distance from A, we find by a similar process that the strain at any point C is

$$= \frac{W . A C}{12 \text{ A } B^2} \times (\text{A } B^3 - \text{A } C^3). \qquad \text{(xii.)}$$

The point of maximum strain in this case is at the distance of $\left(\frac{1}{4}\right)^{\frac{1}{3}}$ A B from A.

PRACTICAL RULES AND EXAMPLES.

RESISTANCE TO CROSS STRAINS.

142. *Prop.* I. To determine a rule for the breadth and depth of a beam, to support a given weight or pressure, when the distance between the supported or strained points is given; when the breadth and depth are both uniformly the same throughout the length, and the strain does not exceed the elastic force of cast iron.

143. *Case* 1. When a beam is supported at the ends, and loaded in the middle, as in fig. 19. From Equation vi. art. 137, taking W for the weight, we have

$$W\, l = \frac{2 f b d^2}{3}, \text{ where } l = \text{F F}'; \text{ fig. 19,}$$

and the value of f is the only part required from experiment; and

$$\frac{3\, l\, W}{2\, b\, d^2} = f.$$

Now in the experiment described in art. 56, Sect. V., the bar returned to its natural state when the load was 300 lbs., and I was perfectly satisfied that it would bear more than that weight without destroying its elastic force. Therefore, from this experiment,

$$\frac{3 \times 34 \times 300}{2} = f = 15300 \text{ lbs.}$$[21]

That is, cast iron of the quality described in art. 56, will bear 15,300 lbs. upon a square inch, when drawn in the direction of its length, without producing permanent alteration in its structure. If this value of f be employed, our equation becomes

$$\frac{3\,l\,W}{2 \times 15300} = b\,d^2;$$

or, as it is convenient to take l in feet,

$$\frac{3 \times 12 \times l \times W}{2 \times 15300} = \frac{l\,W}{850} = b\,d^2.$$

144. *Rule* 1. To find the breadth of an uniform cast-iron beam, to bear a given weight in the middle.

Multiply the length of bearing in feet by the weight to be supported in pounds; and divide the product by 850 times the square of the depth in

---

[21] Mr. Tredgold finds here that cast iron will bear a direct tensile force of 15,300 lbs. per square inch without injury to its elasticity, and concludes (arts. 70-76) that its utmost tensile force is nearly three times as great as this, or upwards of 20 tons. But it will be shown in the "Additions," art. 3, that a less weight per inch than 15,300 lbs. was sufficient to tear asunder bars of several sorts of cast iron; and the mean strength of that metal from experiments on irons obtained from various parts of the United Kingdom did not exceed 16,505 lbs. per square inch. Mr. Tredgold was mistaken in supposing the bar to have borne 300 lbs. without injury to its elasticity, as will be seen under the head 'Transverse Strength' in the Additions.—EDITOR.

inches; the quotient will be the breadth in inches required.[22]

145. *Rule* 2. To find the depth of an uniform cast iron beam, to bear a given weight in the middle.

Multiply the length of bearing in feet by the weight to be supported in pounds, and divide this product by 850 times the breadth in inches; and the square root of the quotient will be the depth in inches.

When no particular breadth or depth is determined by the nature of the situation for which the beam is intended, it will be found sometimes convenient to assign some proportion; as, for example, let the breadth be the nth part of the depth, n representing any number at will. Then the rule becomes—

146. *Rule* 3. Multiply n times the length in feet by the weight in pounds; divide this product by 850, and the cube root of the quotient will be the depth required: and the breadth will be the nth part of the depth.

It may be remarked here, that the rules are the same for inclined as for horizontal beams, when the horizontal distance F F′, fig. 19, is taken for the length of bearing.

---

[22] If the bar is to be of wrought iron, divide by 952 instead of 850.

If the beam be of oak, divide by 212 instead of 850.

If it be of yellow fir, divide by 255 instead of 850.

147. *Example* 1. In a situation where the flexure of a beam is not a material defect, I wish to support a load which cannot exceed 33,600 lbs. (or 15 tons) in the middle of a cast iron beam, the distance of the supports being 20 feet; and making the breadth a fourth part of the depth.

In this case

$$n = 4 \text{ and } \frac{4 \times 20 \times 33600}{850} = 3162{\cdot}35.$$

The cube root of 3162·35 is nearly 14·68 inches, the depth required; the breadth is

$$\frac{14{\cdot}68}{4} = 3{\cdot}87 \text{ inches.}$$

In practice therefore I would use whole numbers, and make the beam 15 inches deep, and 4 inches in breadth.

148. *Case* 2. When a beam is supported at the ends, but the load is not in the middle between the supports. In this case

$$\frac{\mathrm{W.FB} \times \mathrm{F'B}}{l} = \frac{f\,b\,d^2}{6}.$$

(Equation v. art. 136,) consequently

$$\frac{4\,\mathrm{FB} \times \mathrm{F'B} \times \mathrm{W}}{850\,l} = b\,d^2.$$

149. *Rule.* Multiply the distance F B in feet (see fig. 19) by the distance F′ B in feet, and 4 times this product, divided by the whole length F F′ in feet, will give the effective leverage of the load, which being used instead of the length in any of the rules to Case 1, Prob. I., the breadth and depth may be found by them.

150. *Example.* Taking the same example as the last, except that, instead of placing the 15 tons in the middle, it is to be applied at 5 feet from one end; therefore we have F B = 5 feet, and consequently

$$F' B = 15 \text{ feet; and } \frac{5 \times 15 \times 4}{20} = 15$$

the number to be employed instead of the whole length in Rule 3. That is,

$$\frac{4 \times 15 \times 33600}{850} = 2372 \text{ nearly;}$$

and the cube root of 2372 is nearly 13·34 inches, the depth for the beam, and

$$\frac{13{\cdot}34}{4} = 3{\cdot}33 \text{ inches}$$

for the breadth, or nearly $13\frac{1}{2}$ inches by $3\frac{1}{2}$ inches.

In the former case it was 15 inches by 4 inches.

151. *Case* 3. When the load is uniformly distributed over the length of a beam, which is supported at both ends.

In this case

$$\frac{W\,l}{8} = \frac{f\,b\,d^2}{6};$$

(see Equation ix. art. 139,) hence

$$\frac{l\,W}{2 \times 850} = b\,d^2.$$

The same rules apply as in Case 1, art. 144, 145, and 146, by making the divisor twice 850, or 1700.

152. *Example.* In a situation where I cannot make use of an arch for want of abutments, it is necessary to leave an opening 15 feet wide, in an

18-inch brick wall; required the depth of two cast iron beams to support the wall over the opening; each beam to be 2 inches thick, and the height of the wall intended to rest upon the beam being 30 feet?

The wall contains

$$30 \times 15 \times 1\tfrac{1}{2} = 675 \text{ cubic feet};$$

and as a cubic foot of brick-work weighs about 100 ℔s., the weight of the wall will be about 67,500 ℔s.; and half this weight, or 33,750 ℔s., will be the load upon one of the beams. Since the breadth is supposed to be given, the depth will be found by Rule 2, art. 145, if 1700 be used as the constant divisor; thus

$$\frac{15 \times 33750}{1700 \times 2} = 149 \text{ nearly}.$$

The square root of 149 is $12\frac{1}{4}$ nearly; therefore each beam should be $12\frac{1}{4}$ inches deep, and 2 inches in thickness. This operation gives the actual strength necessary to support the wall; but I have usually taken double the calculated weight in practice, to allow for accidents.

In this manner the strength proper for bressummers, lintels, and the like, may be determined. But if there be openings in the wall so placed that a pier rests upon the middle of the length of the beam, then the strength must be found by the rule, art. 145. A rule for a more economical form is given in art. 193.

153. *Case* 4. When a beam is fixed at one end,

and the load applied at the other; also when a beam is supported upon a centre of motion. By Equation x. art. 110,

$$W\,l = \frac{f\,b\,d^2}{6};$$

and taking l in feet, and $f = 15{,}300$ lbs., we obtain

$$\frac{W\,l}{212{\cdot}5} = b\,d^2,$$

but the divisor 212 will be always sufficiently near for practice.

154. *Rule* 1. In a beam fixed at one end, take BD for the length, fig. 17, Plate III., or if the beam be supported in the middle, as in fig. 14, Plate II., take BF or BF′ for the length, observing to use the weight which is to act on that end in the calculation. Then calculate the strength by the rules to Case 1, art. 144, 145, and 146, using

$$\frac{850}{4} = 212$$

instead of 850 as a divisor.

Example. By this rule the proportions for the arms of a balance may be determined. Let the length of the arm, from the centre of suspension to the centre of motion, be $1\frac{1}{2}$ feet; and the extreme weight to be weighed 3 cwt., or 336 lbs., and let the thickness be $\frac{1}{10}$th part of the depth. Then by the rule

$$\frac{10 \times 1{\cdot}5 \times 336}{212} = 24 \text{ nearly.}$$

The cube root of 24 is 2·88 inches, the depth of

the beam at the centre; and the breadth will be 0·288 inch.

For wrought iron the divisor is, in this case, 238, and taking the same example,

$$\frac{10 \times 1{\cdot}5 \times 336}{238} = 21{\cdot}2.$$

The cube root of 21·2 is 2·77 inches, the depth required; and the breadth is 0·277 inch.

155. *Rule* 2. If the weight be uniformly distributed over the length of the beam, employ 425 as a divisor, instead of 850 in the rules to Case 1, art. 144, 145, 146.

156. *Example.* Required the depth for the cantilevers of a balcony to project 4 feet, and to be placed 5 feet apart, the weight of the stone part being 1000 ℔s., the breadth of each cantilever 2 inches, and the greatest possible load that can be collected upon 5 feet in length of the balcony 2200 ℔s.?

Here the weight is

$$1000 + 2200 = 3200 \text{ ℔s.};$$

and by Rule 2, Case 4,

$$\frac{3200 \times 4}{2 \times 425} = 15{\cdot}1 \text{ nearly};$$

and the square root of 15·1 is 3·80 nearly, the depth required.

157. *Remark.* The depth thus determined should be the depth at the wall, as A B, fig. 21, Plate III.; and if the breadth be the same throughout the length, the cantilever will be equally strong in every

part, if the under side be bounded by the straight line BC;[23] therefore, whatever ornamental form may be given to it, it should not be reduced in any part to a less depth than is shown by that line.

158. The strength of the teeth of wheels depends on this case. But since in consequence of irregular action, or any substance getting between the teeth, the whole stress may be thrown upon one corner of a tooth; and it has been shown in art. 111, that the resistance is much less in that case, for then the strength of a tooth of the thickness d would only be

$$\frac{fd^2}{3} = \mathrm{W}.$$

if it were every where of equal thickness; and to make allowance for the diminution of thickness, we ought to make $\frac{fd^2}{5} = \mathrm{W}$. We have also to make an allowance for wear,[24] which will be ample enough at the rate of $\frac{1}{3}$rd of the thickness; therefore,

$$\frac{fd^2 (1 - \frac{1}{3})^2}{5} = \mathrm{W}; \text{ or } \frac{fd^2}{10{\cdot}25} = \mathrm{W}.$$

[23] Emerson's Mechanics, 4to. edit. prop. lxxiii. cor. 2. It was first demonstrated by Galileo, the earliest writer on the resistance of solids. Opere del Galileo, Discorsi, &c., p. 104, tome ii. Bonon, 1655.

[24] The allowance for wear should be for a velocity of 3 feet per second; and in proportion to the velocity, that is,

$$\text{as } 3 : \tfrac{1}{4}\, t :: v : \frac{t\,v}{12}.$$

Hence $\frac{1}{4}\, t - \frac{1}{12}\, t\, v = \frac{1}{4}\,(1 - \frac{1}{3}\, v)\, t$, or $\frac{1}{12}\, t\,(3 - v)$, should be deducted from the thickness in the Table, for velocities differing from 3 feet per second.

In cast iron $f = 15{,}300$; whence we have, with sufficient accuracy,

$$\frac{W}{1500} = d^2, \text{ or } \left(\frac{W}{1500}\right)^{\frac{1}{2}} = d.$$

Rule. Divide the stress at the pitch circle in lbs. by 1500, and the square root of the product is the thickness of the teeth in inches.

Example 1. Let the greatest power acting at the pitch circle of a wheel be 6000 lbs. Then

$$\frac{6000}{1500} = 4;$$

and the square root of 4 is 2 inches, the thickness required.

The breadth of teeth should be proportioned to the stress upon them, and this stress should not exceed 400 lbs. for each inch in breadth, when the pitch[25] is $2\frac{1}{2}$ inches, because the surface of contact is always small, and teeth work irregularly when much worn.

The length of teeth ought not to exceed their thickness, but the strength is not affected by the greater or less length of the teeth.[26]

---

[25] The surface of contact is nearly in the direct ratio of the pitch, and therefore the breadth for a $2\frac{1}{2}$-inch pitch being given, the breadth for any other teeth will be directly as the stress, and inversely as the pitch.

[26] On the length and form of teeth for wheels, the reader may consult the Additions to Buchanan's Essays on Mill-work, vol. i. p. 39, edited by Mr. Rennie, 1842; or the Paper on the Teeth of Wheels, by Professor Willis, given at p. 139 of that work, and in the second volume of the Institution of Civil Engineers.

A Table of the Thickness, Breadth, and Pitch of Teeth for Wheel Work.

Stress in lbs. at the pitch circle.	Thickness of teeth.	Breadth of teeth.	Pitch[27] in inches.
lbs.	inches.	inches.	inches.
400	0·52	1	1·1
800	0·73	2	1·5
1200	0·90	3	1·9
1600	1·03	4	2·2
2000	1·15	5	2·4
2400	1·26	6	2·7
2800	1·36	7	2·9
3200	1·43	8	3·0
3600	1·56	9	3·3
4000	1·64	10	3·4
4400	1·70	11	3·6
4800	1·78	12	3·7
5200	1·86	13	3·9
5600	1·93	14	4·0
6000	2·00	15	4·2

159. As good proportions for the teeth of wheels are of much importance in the construction of machinery, I shall illustrate the mode of applying this Table by examples of different kinds.

Case 1. It is a common mode to compute the stress on the teeth of a machine by the power of the first mover, expressed in horses' power, and the velocity of the pitch circle in feet per second. Now, though I have given the stress in pounds in the Table, I have still kept this popular measure in view; and assuming a horse's power to be 200 lbs.

---

[27] The pitch is the distance from middle to middle of the teeth, and is here made 2·1 times the thickness of the teeth.

with a velocity of 3 feet per second, which we ought to do in calculating the strength of machines,—

Then, the breadth in inches will be equal to the horses' power, to which the teeth are equal, when the velocity of the pitch circle is $1\frac{1}{2}$ feet per second; twice the breadth will be the horses' power when the velocity is 3 feet per second; three times the breadth will be the horses' power when the velocity is $4\frac{1}{2}$ feet per second; four times the breadth will be the horses' power when the velocity is 6 feet per second; five times the breadth will be the horses' power when the velocity is $7\frac{1}{2}$ feet per second; and generally n times the horses' power when the velocity of the pitch circle is n times $1\frac{1}{2}$ feet per second.

Example. Let a steam engine of 10 horses' power be applied to move a machine, and it is required to find the strength for the teeth of a wheel in it, which will move at the rate of 3 feet per second at the pitch circle.

Here then the horses' power should be double the breadth; consequently the breadth will be 5 inches, and, according to the Table, the thickness of the teeth 1·15 inches, and pitch 2·4 inches.

And the same strength of teeth will do for any wheel where the horses' power of the first mover, divided by the velocity in feet per second, produces the same quotient. In this example it is 10 divided by 3; and the same strength of teeth will do for 20 divided by 6; 30 divided by 9; and so on. This will be of some advantage in the arrangement of collections of patterns.

Case 2. When a machine is to be moved by horses, the horses' power should be estimated higher, on account of the jerks and irregular action of horses. We shall not estimate above the strain which often takes place in horse machines, if we rate the horse power at 400 ℔s. with a velocity of 3 feet per second, and make the strength of the teeth accordingly.

But the breadth of the teeth should be made in the same proportion as in the preceding case.

Example. When the horse power is taken at 400 ℔s. with a velocity of 3 feet per second, the stress on the teeth is given for this case in the Table. Thus, in a machine to be moved by four horses, the stress on all the wheels of which the pitch circles move at the rate of 3 feet per second, will be 1600 ℔s., and the pitch should be 2·2 inches, and thickness of teeth 1·03 inches; the breadth half the breadth in the Table, or 2 inches.

Then for any other velocity, as suppose 6 feet per second, it will be, as

$$6 : 3 : : 1600 : 800.$$

That is, the stress on the teeth from a first mover of four horses is 800 ℔s. when the velocity is 6 feet per second; and the thickness of teeth by the Table is 0·73 inch, and pitch 1·5 inches.

Case 3. It remains now to show the general rule which includes the preceding cases, and appears to me to be a more direct and simple mode of proceeding.

If P be the power of the first mover in pounds, and V the velocity of that power in feet per second, the stress on the teeth of a wheel of which the velocity of the pitch circle is v, will be

$$\frac{\mathrm{P\,V}}{v} = \mathrm{W}, \text{ the stress on the teeth.}$$

But we cannot always know the velocity of the pitch circle, because it is not in general possible to vary the number of teeth after the pitch is determined, so as to give it the velocity we have assigned to it before the pitch was known.

The calculation may therefore be made with advantage in this manner: Let N be the number of revolutions the axis is to make per minute, on which the wheel is to be placed; and r the radius the wheel should have if the pitch were two inches, then

$$v = \frac{2{\cdot}1\, d\,\mathrm{N}\, r}{19{\cdot}09 \times 24} = \frac{d\,\mathrm{N}\,r}{218{\cdot}16}.$$

Consequently,

$$\frac{\mathrm{P\,V}}{v} = \frac{218{\cdot}16\,\mathrm{P\,V}}{d\,\mathrm{N}\,r} = \mathrm{W}.$$

Hence

$$\frac{218{\cdot}16\,\mathrm{P\,V}}{1500\,\mathrm{N}\,r} = d^3; \text{ or } \left(\frac{{\cdot}14544\,\mathrm{P\,V}}{\mathrm{N}\,r}\right)^{\frac{1}{3}} = d.$$

The equation affords this rule.

Rule. Multiply 0·146 times the power of the first mover in pounds by its velocity in feet per second, and divide the product by the number of revolutions the wheel is proposed to make per minute, and by the radius the wheel should have in inches if its pitch were 2 inches; the cube root of the quotient will be the thickness of the teeth in inches.

Example 1. Suppose the effective force acting at the circumference of a water-wheel to be 300 lbs. and its velocity 10 feet per second,[28] it is proposed to find the thickness for the teeth of a wheel which is to make twelve revolutions per minute, and have thirty teeth.

Here,

$$0{\cdot}146 \times 300 \times 10 = 438.$$

And since the radius of a wheel with thirty teeth and a pitch of 2 inches is 9·567 inches;[29] we have

$$\frac{438}{12 \times 9{\cdot}567} = 3{\cdot}815.$$

The cube root of 3·815 is very nearly 1·563, the thickness of the teeth required in inches.

Example 2. Let the effective force of the piston of a steam engine be 6875 lbs. and its velocity $3\frac{1}{2}$ feet per second; it is required to determine the

[28] The manner of estimating the effective force, and determining the best velocity for water-wheels, is shown in the Additions to Buchanan's Essays on Mill-work, vol. ii. p. 512-526, second edition; or p. 326-333, in the edit. by Mr. Rennie, 1842. On the subject of water-wheels the reader may consult Mr. Rennie's Preface, page 22, for a notice of the labours of Poncelet, Morin, &c., and the valuable experiments of the Franklin Institute.

[29] This is easily ascertained by Donkin's Table of the radii of wheels. See Buchanan's Essays, vol. i. p. 206, second edition; or p. 114, Rennie's edition.

strength for the teeth of a wheel to be driven by this engine, which is to have 152 teeth, and make 17 revolutions per minute.

In this case, the radius for 152 teeth with a 2-inch pitch is 48·387 inches; therefore,

$$\frac{0{\cdot}146 \times 6875 \times 3{\cdot}5}{17{\cdot}5 \times 48{\cdot}387} = 4{\cdot}15.$$

And the cube root of 4·15 is very nearly 1·6 inches, the thickness of the teeth required.

By referring to the Table it will be found that teeth of this thickness should have a breadth of about 9 inches.

These rules will be found to give proportions extremely near to those adopted by Boulton and Watt, of Soho; Rothwell, Hick, and Rothwell, of Bolton, in Lancashire; and other esteemed manufacturers; which is one of the most gratifying proofs of the confidence that may be placed in the principles of calculation I have followed. The difference is chiefly in the greater breadths I have assigned for the greater strains, and which being a consequence of the principle adopted for proportioning these breadths, I cannot agree to change till it can be shown that the principle is erroneous.

160. *Case* 5. When the pressure upon a beam increases as the distance from one of its points of support. Since the point of greatest strain is at $\sqrt{\frac{1}{3}}\, l$ from the point A, where the strain begins at, (see fig. 20,) we have by art. 140 and 110,

$$\frac{W\, l}{7{\cdot}75} = \frac{f\, b\, d^2}{6},$$

or when l is in feet, and

$$f = 15300 \text{ lbs.}; \ \frac{W\,l}{1647} = b\,d^2;$$

a result which differs so little from Case 3, that the same rule may serve for both cases.

161. *Prop.* II. To determine a rule for the diagonal of an uniform square beam to support a given strain in the direction of that diagonal; when the strain does not exceed the elastic force of cast iron.

162. *Case* 1. When a beam is supported at the ends and loaded in the middle,

$$\frac{W\,l}{4} = \frac{f\,a^3}{24},$$

art. 137 and 112; or when l is in feet, and

$$f = 15300 \text{ lbs.} \left(\frac{W\,l}{212}\right)^{\frac{1}{3}} = a.$$

163. *Rule.* Multiply the length in feet by the weight in pounds, and divide the product by 212; the cube root of the quotient is the diagonal of the beam in inches.

164. *Case* 2. When a beam is supported at the ends, and the strain is not in the middle of the length,

$$\frac{W \times FB \times F'B}{l} = \frac{f\,a^3}{24},$$

art. 112 and 136; or when $f = 15{,}300$ lbs. and the length and distances F B, F′ B from the ends are in feet,

$$\left(\frac{W \times FB \times F'B}{53\,l}\right)^{\frac{1}{3}} = a.$$

165. *Rule.* Multiply the weight in pounds by

the distance F B in feet, and multiply this product by the distance from the other end, or F′ B in feet (see fig. 19). Divide the last product by 53 times the length, and the cube root of the quotient will be the diagonal of the beam in inches.

I limit the rules to these cases only, because a beam is seldom placed in the position described in this proposition. Examples are omitted for the same reason.

166. *Prop.* III. To determine a rule to find the diameter of a solid cylinder, to support a given strain, when the strain does not exceed the elastic force of cast iron.

If the diameter be not uniform, the diameter determined by the rule will be that at the point of greatest strain, and the diameter at any other point should never be less than corresponds to the form of equal strength.

167. *Case* 1. When a solid cylinder is supported at the ends, and the weight acts at the middle of the length,

$$\frac{W\,l}{4} = \frac{\cdot 7854\, f\, d^3}{8},$$

art. 113 and 137; or when l is in feet,

$f = 15300$ lbs. and $d =$ the diameter in inches,

we have

$$\left(\frac{W\,l}{500}\right)^{\frac{1}{3}} = d.$$

168. *Rule.* Multiply the weight in pounds by the length in feet; divide this product by 500, and

the cube root of the quotient will be the diameter in inches.[30]

The figure of equal strength for a solid, of which the cross section is every where circular, is that generated by two cubic parabolas, set base to base,[31] the bases being equal, and joining at the section where the strain is the greatest.

169. *Example.* Required the diameter of a horizontal shaft of cast iron to sustain a pressure of 2000 ℔s. in the middle of its length; the length being 20 feet? In this case we have

$$\frac{2000 \times 20}{500} = 80;$$

and the cube root of 80 is 4·31 inches nearly, which is the diameter required.

This is supposed to be a case where the flexure is of no importance, otherwise the diameter must be determined by the rules for flexure.

170. *Case* 2. When a cylinder is supported at the ends, but the strain is not in the middle of the length. By art. 113 and 136,

$$\frac{\text{W} \times \text{FB} \times \text{F}'\text{B}}{l} = \frac{\cdot 7854\, f\, d^3}{8};$$

or when the lengths are in feet, d is the diameter in inches, and $f = 15{,}300$, the equation becomes

$$\left(\frac{4\,\text{W} \times \text{FB} \times \text{F}'\text{B}}{500\, l}\right)^{\frac{1}{3}} = d.$$

---

[30] For wrought iron divide by 560 instead of 500. For oak divide by 125 instead of 500.

[31] Emerson's Mechanics, 4to. edit. prop. lxxiii. cor. 4.

171. *Rule.* Multiply the rectangle of the segments, into which the strained point divides the beam, in feet, by 4 times the weight in pounds; when this product is divided by 500 times the length in feet, the cube root of the quotient will be the diameter of the cylinder in inches.

The figure of equal strength is the same as in Case 1, art. 168.

172. *Example.* Required the diameter of a shaft of cast iron to resist a pressure of 4000 ℔s. at 3 feet from the end, the whole length of the shaft being 14 feet? In this example

$$\frac{3 \times 11 \times 4 \times 4000}{500 \times 14} = 75{\cdot}43.$$

The cube root of 75·43 is nearly 4·23 inches, the diameter required.

173. *Case* 3. When a load is uniformly distributed over the length of a solid cylinder supported at the ends only. By art. 113 and 139,

$$\frac{W\,l}{8} = {\cdot}7854\, f\, d^3;$$

therefore, when l is in feet, d the diameter in inches, and $f = 15{,}300$, we have

$$\left(\frac{W\,l}{1000}\right)^{\frac{1}{3}} = d = \tfrac{1}{10}\,(W\,l)^{\frac{1}{3}}.$$

174. *Rule.* Multiply the length in feet by the weight in pounds, and $\frac{1}{10}$th of the cube root of the product will be the diameter in inches.[32]

---

[32] For wrought iron divide by 10·38.

The figure of equal strength for an uniform load, the section being every where circular,[33] is that generated by the revolution of a curve of which the equation is

$$a\,(l\,x - x^2)^{\frac{1}{3}} = y.$$

175. *Example.* A load of 6 tons (or 13,440 ℔s.) is to be uniformly distributed over the length of a solid cylinder of cast iron, of which the length is 12 feet; required its diameter, so that the load shall not exceed its elastic force?

In this case

$$12 \times 13440 = 161280;$$

and the cube root of 161,280 is 54·44, and $\frac{1}{10}$th of this is 5·444 inches, the diameter required.

176. *Case* 4. When a cylinder is fixed at one end, and the load applied at the other; also, when a cylinder is supported on a centre of motion. By art. 113,

$$\mathrm{W}\,l = \cdot 7854\,f\,r^3;$$

therefore, when d is the diameter, l is in feet, and $f = 15{,}300$ ℔s., we have

$$\left(\frac{8\,\mathrm{W}\,l}{1000}\right)^{\frac{1}{3}} = d, \text{ or } \tfrac{1}{5}\,(\mathrm{W}\,l)^{\frac{1}{3}} = d.$$

The figure of equal strength is the same as in Case 1, art. 168.

177. *Rule.* Multiply the leverage the weight acts with, in feet, by the weight in pounds; the fifth

---

[33] Emerson's Mechanics, prop. lxxiii. cor. 3.

part of the cube root of this product will be the diameter required in inches.

The most important application of this case is to determine the proportions for gudgeons and axles; and this application will be best illustrated by an example.

The greatest stress upon a gudgeon or axle takes place when, from any accident, that stress is thrown upon the extreme point of its bearing. But besides the greatest possible stress we have to provide for wear; perhaps $\frac{1}{5}$th of the diameter may be allowed for this purpose.

Now taking the length l for the length from the shoulder to the extreme point of bearing in inches, we have

$$\tfrac{1}{5}\left(\tfrac{1}{12}\, l\, W\right)^{\frac{1}{3}} = d\left(1 - \tfrac{1}{5}\right); \text{ or } \tfrac{1}{9}\left(l\, W\right)^{\frac{1}{3}} = d.$$

Whence we have this practical rule: Multiply the stress in pounds by the length in inches, and the cube root of the product divided by 9 is the diameter of the gudgeon in inches.[34]

Example. Let the stress on the gudgeon be 10 tons, or 22,400 lbs., and its length 7 inches. Then

$$7 \times 22400 = 156800;$$

---

[34] For wrought iron divide by 9·34. For wheel carriages less than 3-inch axles the length may be 5 times the diameter; then for wrought iron,

$$\frac{\sqrt{W}}{13} = d.$$

Above 3 inches it may be 4 times the diameter.

and the cube root of this number is 54 nearly; and

$$\frac{54}{9} = 6 \text{ inches},$$

the diameter required.

But the stress of a gudgeon on its bearings ought to be limited, otherwise they will wear away very quickly: let us suppose this stress to be confined to a portion of the circumference, which is equal to $\frac{3}{4}$ths of the diameter of the gudgeon; and that the pressure is limited to 1500 lbs. upon a square inch, which is about as great a pressure as we ought to put on the rubbing surfaces when one of them is of gun-metal. In this case we shall have

$$l = \frac{4 \text{ W}}{3 \times 15000 \, d}; \text{ or } l = \frac{\text{W}}{1125 \, d};$$

and to allow for a small portion of freedom we make

$$l = \frac{\text{W}}{1000 \, d}.$$

If this value of l be introduced in the preceding equation, we have

$$\frac{1}{90}\left(\frac{\text{W}^2}{d}\right)^{\frac{1}{3}} = d; \text{ or W} = 854 \, d^2; \text{ and } l = \cdot 854 \, d.$$

According to these principles the following Table has been calculated, and I hope it will be useful.

Table of the Proportions of Gudgeons and Axles for different degrees of Stress.

Diameter of gudgeons.	Length of gudgeons.	Stress they may sustain.
inches.	inches.	lbs.
$\frac{1}{2}$	·43	213
$\frac{3}{4}$	·64	480
1	·85	854
$1\frac{1}{2}$	1·25	1,921
2	1·7	3,416
3	2·5	7,686
4	3·4	13,664
5	4·3	21,350
6	5·1	30,744
7	5·9	41,846
8	6·8	54,656
9	7·7	69,174
10	8·5	85,400

Gudgeons exposed to the action of gritty matter may be made larger in diameter about $\frac{1}{8}$th part.

178. *Prop.* IV. To determine a rule for the exterior diameter of an uniform tube or hollow cylinder[35] to resist a given force where the strain does not exceed the elastic force of cast iron.

179. *Case* 1. When a tube is supported at the ends, and the load acts at the middle of the length. By art. 115 and 137,

$$\frac{W\,l}{4} = \cdot 7854\, f\, d^3\, (1 - N^4);$$

[35] A considerable accession of strength and stiffness is gained by making shafts hollow, which has been illustrated in art. 115; but it is difficult to get them cast sound, therefore shafts of this kind require to be carefully proved.

hence, when d is the diameter in inches, l the length in feet, and $f = 15{,}300$ lbs., we have

$$\left(\frac{\mathrm{W}\,l}{500\,(1-\mathrm{N}^4)}\right)^{\frac{1}{3}} = d.$$

180. *General Rule.* Fix on some proportion between the diameters; so that the exterior diameter is to the interior diameter as 1 is to N; the number N will always be a decimal, and ought not to exceed 0·8. [36]

Then multiply the length in feet by the weight to be supported in pounds. Also, multiply 500 by the difference between 1 and the fourth power of N, and divide the product of the length and the weight by the last product, and the cube root of the quotient will be the diameter in inches.

The interior diameter will be the number N multiplied by the exterior diameter, and half the difference of the diameters will be the thickness of metal.

If the proportion between the exterior and interior diameter be fixed, so that the thickness of

---

[36] In a large shaft there should be a tolerable bulk of metal to secure a perfect casting. Mr. Buchanan, in his 'Essay on the Shafts of Mills,' vol. i. p. 305, second edition, (or page 202-3 in the edition of 1841,) describes a hollow shaft of which the exterior diameter was 16 inches, and the interior one 12 inches, therefore

$$16 : 12 :: 1 : \mathrm{N} = \frac{12}{16} = \cdot 75.$$

This shaft was for an over-shot water-wheel of 16 feet diameter.

metal may be always $\frac{1}{5}$th of the exterior diameter of the tube; then N = ·6; and the rule is

$$\left(\frac{W\,l}{435}\right)^{\frac{1}{3}} = d.$$

And there being no difference between this equation and that for a solid cylinder, except the constant divisor, we have this rule:

Particular Rule. When the thickness of metal is to be $\frac{1}{5}$th of the diameter of the tube, let the diameter be calculated by the rule for a solid cylinder, art. 168, except that 435 is to be used as a divisor instead of 500.

181. *Example.* Let the weight of a water-wheel, including the weight of the water in the buckets, be 44,800 lbs., and the whole length of the shaft 8 feet; from which deducting 5 feet,[37] the width of the wheel, leaves 3 feet for the length of bearing: required the diameter of a hollow shaft for it?

Making N = ·7, its fourth power is ·2401; and

$$1 - \cdot 240 = \cdot 76.$$

Therefore, by the general rule we have

$$\frac{3 \times 44800}{500 \times \cdot 76} = 354 \text{ in the nearest whole numbers;}$$

and the cube root of 354 is 7 inches, the exterior diameter; and

$$7 \times \cdot 7 = 4 \cdot 9 \text{ inches, the interior diameter.}$$

By the particular rule the computation is easier, for it is

---

[37] The wheel being so framed that the part of the length of the shaft it occupies may be considered perfectly strong.

$$\frac{3 \times 44800}{435} = 309;$$

and the cube root of 309 is 6·76 inches, the exterior diameter; and the thickness of metal $\frac{1}{5}$th of this, or $1\frac{3}{8}$ inches nearly.

The particular rule will be found to give a good proportion for the thickness of metal for considerable strains; but in lighter work, where stiffness is the chief object, recourse should be had to the general rule.

182. *Case* 2. When a tube is supported at the ends, but the strain is not in the middle of the length. When the necessary substitutions are made, we have, by art. 115 and 136,

$$\left(\frac{4\,W \times FB \times F'B}{500\,l \times (1 - N^4)}\right)^{\frac{1}{3}} = d.$$

183. *Rule.* Multiply the rectangle of the segments into which the strained point divides the beam, in feet, by four times the weight in pounds; call this the first product.

Multiply 500 times the length, in feet, by the difference between 1 and the fourth power of N; (N being the interior diameter when the exterior diameter is unity;) call this the second product.

Divide the first product by the second, and the cube root of the quotient will be the exterior diameter of the tube in inches.

Or, making the thickness of metal $\frac{1}{5}$th of the diameter, calculate by the rule art. 171, using 435 instead of 500 as a divisor.

184. *Example.* Let the weight of a wheel and other pressure upon a shaft be equal to 36,000 lbs., the distance of the point of stress from the bearing at one end being 3 feet, and the distance from the other bearing 1·5 feet; N being ·8; required the exterior and interior diameter of the shaft?

The fourth power of ·8 is ·409, and

$$1 - \cdot 409 = \cdot 591.$$

Therefore by the rule

$$\frac{3 \times 1{\cdot}5 \times 4 \times 36000}{500 \times 4{\cdot}5 \times \cdot 591} = 485;$$

and the cube root of 485 is 7·86 inches, the exterior diameter, and

$$7{\cdot}86 \times \cdot 8 = 6{\cdot}3 \text{ inches, the interior diameter.}$$

Cases 3 and 4 are not likely to occur in the practical application of tubes, but they may be supplied by Cases 3 and 4 for solid cylinders, by dividing the diameter of the solid cylinder by the cube root of the difference between 1 and the fourth power of N; or when the thickness of metal is to be $\frac{1}{5}$th of the diameter, divide by 435 instead of 500.

185. *Prop.* v. To determine a rule for finding the depth of a beam of the form of section shown in fig. 9, Plate I., to resist a given force when the strain does not exceed the elastic force of cast iron.

186. *Case* 1. When the beam is supported at the ends, and the load acts in the middle of the length. By art. 116 and 137,

$$\frac{W\,l}{4} = \frac{f\,b\,d^2}{6}(1 - q\,p^3);$$

or making l = the length in feet, and f = 15,300 lbs.,

$$\frac{W\, l}{850} = b\, d^2\, (1 - q\, p^3).$$ [38]

187. *Rule.* Assume a breadth $a\,b$, fig. 9, that will answer the purpose the beam is intended for; and let this breadth, multiplied by some decimal q, be equal to the sum of the projecting parts, or, which is the same thing, equal to the difference between the breadth of the middle part and the whole breadth.

Also, let p be some decimal which multiplied by the whole depth will give the depth of the middle or thinner part $e\,f$ in the figure.

Multiply the length in feet by the weight in pounds, and divide this product by 850 times the breadth multiplied into the difference between unity and the cube of p multiplied by q; the square root of the quotient will be the depth in inches.

---

[38] If we make $p = \cdot 7$, and $q = \cdot 6$; then,

$$850\, (1 - q\, p^3) = 675;$$

and the rule is

$$\frac{W\, l}{675} = b\, d^2;$$

and the breadth of the middle part = $\cdot 4\, b$, and the depth of the middle part $\cdot 7\, d$.

When the parts are in these proportions, the strength is to that of the circumscribed rectangular section as 1 : 1·26.

If, with the same proportions, we make the breadth $a\,b$ always one-fifth of the depth $b\,d$, fig. 9, the strength will be to that of a square beam of the same depth as 1 : 6·3; and the stiffness will be in the same proportion.

The figure of equal strength for this case is formed by two common parabolas put base to base, as shown by the dotted lines in fig. 22; for $l : d^2$ a property of the parabola, the other being constant quantities. Fig. 22 shows how it may be modified to answer in practice. When a figure of equal strength is used, the depth determined by the rule is that at the point of greatest strain, as C D in the figure.

188. *Example* 1. Required the depth of a beam of cast iron of the form of section shown in fig. 9, Plate I., to bear a load of 33,600 lbs. in the middle of the length, the length being 20 feet, and the breadth, *a b*, 3 inches?

Take ·625 for the decimal q, and ·7 for the decimal p, which are proportions that will be found to answer very well in practice.[39]

Then

$$\frac{20 \times 33600}{850 \times 3 \times (1 - \cdot 625 \times \cdot 7^3)} = \frac{20 \times 33600}{3 \times 667} = 335 \cdot 4 \text{ nearly};$$

and the square root of 335·4 is 18·4 inches, the depth required.

The depth *b d* being 18·4, the depth *e f* will be

$$18 \cdot 4 \times \cdot 7 = 12 \cdot 88 \text{ inches; also,}$$

$$3 \times \cdot 625 = 1 \cdot 875, \text{ and}$$

---

[39] Since

$$850\ (1 - \cdot 625 \times \cdot 7^3) = 677 \text{ nearly};$$

whenever the same proportions are used, the divisor 677 may be employed instead of repeating the calculation.

$$3 - 1{\cdot}875 = 1{\cdot}125 \text{ inches},$$

the breadth of the middle part of the section.

Comparing this with the example, art. 147, it will be found that the same weight requires only about $\frac{2}{3}$rds of the quantity of iron to support it, when the beam is formed in this manner.

Example 2. The same rule applies to determining the size of the rails for an iron railway, where economy with strength and durability is of much importance. As the weight has to move over the length of the rail, the figure of equal strength is that shown in Plate III. fig. 24, only it should be placed with the straight side upwards.

Suppose the weight of a coal waggon to be about 4 tons, 8960 lbs.; when the rails are shorter than twice the distance between the wheels, the utmost strain on a rail cannot exceed half this weight, or 4480 lbs., which will be allowing half the strength nearly for accidents. The usual length of one rail is 3 feet,[40] and supposing the breadth to be 2 inches,

---

[40] It is worthy of consideration whether this be the most economical length, or not, for rails. This may be done as follows:

The weight of a bar of iron, an inch square and 700 feet long, is 1 ton; therefore, for a length of 700 feet, the area of the bar in inches multiplied by the price of a ton of iron will be the amount of 700 feet of rail. Make $\frac{700}{x}$ the length of a single rail; then, supposing the rail all of the same thickness,

$$\sqrt{\frac{W \times 700}{850 \times b\,x}} = d,$$

then, by the manner of calculation shown in the note to art. 186,

$$\frac{W\,l}{675 \times b} = \frac{4480 \times 3}{675 \times 2} = d^2 = 9{\cdot}96\,;$$

and the square root of

---

the depth, and when it is reduced at the ends,

$$\cdot 7\, b \sqrt{\frac{W \times 700}{850 \times b\,x}} = \text{the area}:$$

and calling A the price of a ton of iron; and B the price of fixing, and materials for one block; then the price of 700 feet will be

$$\cdot 7\, A\, b \sqrt{\frac{W \times 700}{850\, b\, x}} + x\, B = \frac{\cdot 64\, A \sqrt{W\, b}}{\sqrt{x}} + B\, x.$$

Hence by the rules of maxima and minima it appears that the price will be the least when the number of supports for 700 feet is

$$= \left(\frac{\cdot 32\, A \sqrt{W\, b}}{B}\right)^{\frac{2}{3}};$$

wherein W is half the weight of a waggon and its load in lbs.

The same equation will apply to the new railway invented by Mr. Palmer, when W is made the whole weight of the waggon in lbs.

An example will illustrate the application: Let A the price of a ton of iron be £8; B the price of one support £0·5; the weight of a waggon 8960 lbs.; and the breadth of the rail 3 inches. Then

$$\left(\frac{\cdot 32 \times 8 \times \sqrt{8960 \times 3}}{\cdot 5}\right)^{\frac{2}{3}} = 89\,;$$

that is, there should be 89 supports in 700 feet, in order that the expense may be the least possible at these prices, and for these proportions; which makes the distance of the supports nearly 8 feet. But it should be understood that these prices are only what I have inserted for illustration; they are not from actual estimate.

$$9{\cdot}96 = 3{\cdot}16 \text{ inches},$$

the depth in the middle of the length.

Also,

$$3{\cdot}16 \times {\cdot}7 = 2{\cdot}212 \text{ inches},$$

the depth of the thin part in the section at the middle of the length, and

$$2 \times {\cdot}4 = 0{\cdot}8 \text{ inch},$$

the thickness of the middle part of the section.

The depth of a rail, all of the same thickness, would be 2·83 inches in the middle, calculated by Rule 2, art. 145.

Example 3. In Palmer's railway a single rail carries the waggon;[41] and let its weight be 8960 lbs., and the length of the rail 8 feet, its breadth 3 inches. By the rule

$$\frac{W\,l}{675\,b} = \frac{8960 \times 8}{675 \times 3} = 35{\cdot}4.$$

The square root of 35·4 is very nearly 6 inches, the depth required; and the depth of the middle part

$$6 \times {\cdot}7 = 4{\cdot}2 \text{ inches}.$$

The breadth 3 inches, and breadth of middle part

$$3 \times {\cdot}4 = 1{\cdot}2 \text{ inches}.$$

These are the dimensions for the middle of the length; but the under edge should be the figure of equal strength, Plate III. fig. 24, with the straight side upwards.

---

[41] Description of a Railway on a New Principle, by H. R. Palmer, 8vo. London, 1823.

189. *Case* 2. When the beam is supported at the ends, but the load not applied in the middle between the supports. When l is the length in feet, and $f = 15{,}300$ lbs.,

$$\frac{4\ \mathrm{F\,B} \times \mathrm{F'\,B} \times \mathrm{W}}{850\ l\ (1 - p^3\ q)} = b\ d^2$$

by art. 116 and 136.

190. *Rule.* Multiply the rectangle of the segments into which the strained point divides the beam, in feet, by 4, and divide this product by the length in feet; use this quotient instead of the length of the beam, and proceed by the last rule.

191. *Example.* Let the load to be supported be 33,600 lbs. at 5 feet from one end, the whole length being 20 feet. Also, let the breadth of the widest part $a\,b$, fig. 9, be 4 inches.

Here F B = 5 feet, therefore F′ B = 15 feet, and

$$\frac{4 \times 5 \times 15}{20} = 15:$$

the multiplier to be used instead of the whole length in the rule.

Let $p = \cdot 7$, and $q = \cdot 625$; then

$$\frac{15 \times 33600}{850 \times 4 \times (1 - \cdot 625 \times \cdot 7^3)} = \frac{15 \times 33600}{4 \times 677} = 189 \text{ nearly,}$$

of which the square root is 13·5 inches, the depth required.

The depth $e\,f$ will be

$$\cdot 7 \times 13 \cdot 5 = 9 \cdot 45 \text{ inches,}$$

and the breadth of the middle part of the section will be

$$4 - \overline{4 \times \cdot 625} = 4 - 2\cdot 5 = 1\cdot 5 \text{ inches.}$$

192. *Case* 3. When the load is uniformly distributed over the length of a beam. In this case

$$\frac{\mathrm{W}\, l}{1700\,(1 - q\, p^3)} = b\, d^2$$

by art. 116 and 139.

193. *Rule.* Use half the weight instead of the whole weight upon the beam, and proceed by the rule to Case 1, art. 187.

The form of equal strength for this case, when the breadth is uniform, is an ellipse, but in practical cases it will require to be altered to the form shown in fig. 24.

194. I propose to give as an example of this rule, its application to the construction of fire-proof buildings; but it also applies to rafters, girders, bressummers, and all cases where the load is uniformly distributed over the length.

A fire-proof floor is usually formed by placing parallel beams of cast iron across the area in the shortest direction, and arching between the beams as shown by fig. 10, Plate I., with brick or other suitable material. Or they may be done by flat plates of iron resting on the ledges, with one or two courses of bricks paved upon the iron plates; and when the distance of the joists is considerable, the iron plates may be strengthened by ribs on the upper side as the floor plates of iron bridges are made.

When arches are employed, floors of this kind

are least expensive when the arches are of considerable span; but then it is necessary to provide against the lateral thrust of the arches by tie bars. Also, since the arches ought to be flat, we can only extend them to a limited span, otherwise they would be too weak to answer the purpose. For instance, when an arch is to rise only $\frac{1}{10}$th of the span, and to be half a brick or 4½ inches thick,[42] the greatest span that can be given to the arch with safety in a floor for ordinary purposes is 5 feet. If the arch rise only $\frac{1}{12}$th of the span, the span must be limited to 4 feet; and if it rise only $\frac{1}{17}$th of the span, it must be limited to 3 feet.

Again, for arches of one brick, or 9 inches, to bear the same load, and the rise $\frac{1}{10}$th of the span, the greatest span that can be given with safety is 8 feet;[43] when the rise is $\frac{1}{12}$th of the span, 7 feet; and when the rise is only $\frac{1}{17}$th of the span, the greatest span should not exceed 5 feet.

These limits were calculated from the ordinary strength of brick, and on the supposition that the load upon the floor will never be greater than 170 lbs. upon a superficial foot, in addition to the weight of the floor itself. If the load be greater, the span must be less, or the rise greater.[44]

For half brick arches the breadth of the beam *c d*,

---

[42] Rad. of curv. 6·75 feet.

[43] Rad. of curv. 15·6 feet.

[44] See also Elementary Principles of Carpentry, art. 249 and 270; edition by Mr. Barlow, 1840.

fig. 9, should be about 2 inches; and for 9-inch arches, from $2\frac{1}{2}$ to 3 inches.

Example. It is proposed to form a fire-proof room, but from its situation it cannot be vaulted in the ordinary way on account of the strong abutments required for common vaulting, and also common vaulting is objectionable, because so much space is lost in a low room. The shortest direction across the room is 12 feet, and if iron beams of 3 inches breadth be laid across at 5 feet apart, and arched between with 9-inch brick arches, it is required to find the depth for the beams? See fig. 10, Plate I.

The quantity of brick work resting upon 1 foot in length of joist will be

$$5 \times \cdot 75 = 3 \cdot 75 \text{ cubic feet;}$$

and the weight of a cubic foot being nearly 100 lbs., the weight of the brick work will be 375 lbs.

But since the space above is to be used; and the greatest probable extraneous weight that will be in the room will arise from its being filled with people, we may take that weight at 120 lbs. per superficial foot, and we have

$$5 \times 120 = 600 \text{ lbs.}$$

for the weight on 1 foot in length. And supposing the paving and iron to be 350 lbs. for each foot in length; the whole load on a foot in length will be

$$375 + 600 + 350 = 1325 \text{ lbs. or}$$

$$12 \times 1325 = 15900 \text{ lbs.}$$

the whole weight upon one joist. And as half this weight multiplied by the length, and divided by the breadth and constant number,[45] is equal to the square of the depth, we have

$$\frac{7950 \times 12}{675 \times 3} = 47{\cdot}11,$$

of which the square root is nearly 7 inches, the depth required. And

$$7 \times {\cdot}7 = 4{\cdot}9 \text{ inches}$$

the depth of the middle part, and

$$3 \times {\cdot}4 = 1{\cdot}2$$

the breadth of the middle part.

By fixing the breadth, you avoid the risk of calculating for a thinner beam than is sufficient to support firmly the abutting course of bricks.

By means of this example we may easily form a small Table of the depth of beams for fire-proof floors, which will be often useful: in so doing, I shall not regard the difference between the weight of a 9-inch and a 4½-inch floor; because the lighter floor will be more liable to accidents from percussion, and therefore should have excess of strength.

---

[45] See note to Rule, art. 186.

Table of Cast Iron Joists for Fire-proof Floors, when the extraneous load is not greater than 120 lbs. *on a superficial foot* (*see* FLOORS, *Alphabetical Table*).

Length of joists in feet.	Half brick arches, breadth of beams 2 inches.			Nine-inch arches, breadth of beams 3 inches.		
	3 feet span.	4 feet span.	5 feet span.	6 feet span.	7 feet span.	8 feet span.
feet.	Depth in inches.	Depth in inches.	Depth in inches.	Depth in inches.	Depth in inches.	Depth in inches.
8	$4\frac{1}{2}$	$5\frac{1}{4}$	$5\frac{3}{4}$	$5\frac{1}{4}$	$5\frac{3}{4}$	6
10	$5\frac{1}{2}$	$6\frac{1}{2}$	7	$6\frac{1}{2}$	$7\frac{1}{4}$	$7\frac{1}{2}$
12	$6\frac{3}{4}$	$7\frac{3}{4}$	$8\frac{1}{2}$	$7\frac{3}{4}$	$8\frac{1}{2}$	9
14	$7\frac{3}{4}$	9	10	$9\frac{1}{4}$	10	$10\frac{1}{2}$
16	9	$10\frac{1}{4}$	$11\frac{1}{4}$	$10\frac{1}{2}$	$11\frac{1}{2}$	12
18	10	$11\frac{3}{4}$	$12\frac{3}{4}$	$11\frac{3}{4}$	13	$13\frac{1}{2}$
20	$11\frac{1}{4}$	13	14	13	$14\frac{1}{4}$	15
22	$12\frac{1}{4}$	$14\frac{1}{4}$	$15\frac{1}{2}$	$14\frac{1}{4}$	$15\frac{3}{4}$	$16\frac{1}{2}$
24	$13\frac{1}{4}$	$15\frac{1}{2}$	17	$15\frac{1}{2}$	17	18

For half brick arches the breadth *a b*, fig. 9, Plate I., is to be 2 inches, and the thickness of the middle part $\frac{8}{10}$ths of an inch; the depth *e f* being $\frac{7}{10}$ths of the whole depth; and the whole depth is given in inches in the Table for each length and span.

For 9-inch arches the breadth *a b*, fig. 9, is to be 3 inches, and the breadth of the middle part 1 inch and $\frac{2}{10}$ths. The depth $\frac{7}{10}$ths of the whole depth, as in the $4\frac{1}{2}$-inch arches.

If the floor be for a room of greater span than about 16 feet, let the beams be put 8 feet apart; and put the beams for 8 feet bearing across at right angles to the other, in the manner of binding joists, and arch between the shorter beams. By casting the shorter beams with flanches at the ends, they can be bolted to the other, and a complete firm floor be made. This method has also the advantage of rendering it extremely easy to fix either a wooden floor or a ceiling.

The construction of these floors renders a place secure from fire without loss of space, and with very little extra expense; it may be of infinite use in the preservation of deeds, libraries, and indeed of every other species of property. In a public museum, devoted to the collection and preservation of the scattered fragments of literature and art, it is extremely desirable that they should be guarded against fire; otherwise they may be involved in one common ruin, more dreadful to contemplate than their widest dispersion.

195. *Case* 4. When a beam is fixed at one end, and the load applied at the other. Also, when a beam is supported upon a centre of motion. By art. 116,

$$W l = \frac{f b d^2}{6} \times (1 - p^3 q),$$

or when l is in feet, and $f = 15{,}300$ lbs.,

$$\frac{W l}{212 (1 - q p^3)} = b d^2.$$

196. *Rule* 1. Calculate by the rule to Case 1, art. 187, using 212 instead of 850 for a divisor.

Or when the breadth of the middle is made $\frac{4}{10}$ths of the extreme breadth, and the depth ef in fig. 9 is $\frac{7}{10}$ths of the whole depth; then, calculate by the rules art. 144, 145, or 146, using 168 instead of 850 as a divisor.

The figure of equal strength is a parabola; see figs. 25 and 26.

197. *Rule* 2. If the weight be uniformly distributed over the length, take the whole load upon the beam for the weight, and calculate by the rule to Case 1, art. 187, except using 425 instead of 850 as a divisor.

198. *Prop.* VI. To determine a rule for finding the depth of a beam when part of the middle is left open, as in figs. 11, 12, and 27, so that it will resist a given force; the strain not exceeding the elastic force of the material.

199. When the depth is more than 12 or 14 inches, angular parts in the middle become necessary, as in fig. 27; the disposition of the middle part may in a great measure be regulated by fancy, provided it allows of sufficient diagonal and cross ties to bind the upper and lower parts together. The middle parts should be made of the same size as the other, in order that they may not be rendered useless by irregular contraction.

If the beams be required so long as not to be made in a single casting, and it is not a good plan to cast in very long lengths, then they may be joined in the middle, as in fig. 27. The connexion is made at the lower side only; at the upper side let the parts abut against one another, with only some contrivance to steady them while they become fixed in their places and loaded. Fig. 28 is a plan of the under side, showing how the connexion may be made.

200. *Case* 1. When the beam is supported at the ends, and the load acts at the middle of the length. By art. 117 and 137,

$$\frac{W\,l}{4} = \frac{f\,b\,d^2}{6}(1 - p^3),$$

or making l = the length in feet, and f = 15,300 lbs.,

$$\frac{W\,l}{850\,(1 - p^3)} = b\,d^2.$$

Now, in general, we may make $p = \cdot 7$, and then,

$$\frac{W\,l}{558} = b\,d^2;$$

or, in practice,[46]

$$\frac{W\,l}{560} = b\,d^2.$$

$$\text{If } b = \frac{1}{9}\,d, \text{ then } \frac{W\,l}{62} = d^3.$$

201. *Rule.* Multiply the length in feet by the weight to be supported in pounds; and divide this

---

[46] If we make $p = 0{\cdot}6$, then

$$\frac{W\,l}{135} = d^3; \text{ and } b = 0{\cdot}2\,d,$$

and the depth of the section at A B or C D, fig. 11, Plate II., will be the same as the breadth of the beam. And as the equation for a square beam of the same depth is

$$\frac{W\,l}{850} = d^3,$$

the strength of this beam will be to that of the square beam, of the same depth, as 1 : 6·3.

product by 560 times the breadth in inches; the square root of the quotient will be the depth required in inches. Consult art. 41 and 43 respecting the form of beams of this kind. The depth between the upper and lower part of the beam will be ·7 *d* inch, where *d* is the depth found by the rule.

202. *Example.* A beam for a 30-feet bearing is intended to sustain a load of 6 tons (13,440 lbs.) in the middle, the breadth to be 4 inches; required the depth?

By the rule

$$\frac{30 \times 13440}{4 \times 560} = 180;$$

the square root of 180 is nearly 13·5 inches, the whole depth.

The depth between the upper and lower part is

$$\cdot 7 \times 13\cdot 5 = 9\cdot 45 \text{ inches.}$$

If the depth be given, suppose 16 inches, and the breadth be required, then

$$\frac{30 \times 13440}{16 \times 16 \times 560} = 2\cdot 82, \text{ the breadth in inches;}$$

when the depth is 16 inches, and the depth between the upper and lower parts is

$$\cdot 7 \times 16 = 11\cdot 2 \text{ inches.}$$

203. *Case* 2. When a beam is supported at the ends, but the load is not applied at the middle.

When l is the length in feet, $p = \cdot 7$, and $f =$ 15,300 lbs.,

$$\frac{4\,\mathrm{BC} \times \mathrm{CD} \times \mathrm{W}}{558\,l} = b\,d^2,$$

(see fig. 12, Plate II. ;) or

$$\frac{BC \times CD \times W}{139\,l} = b\,d^2.$$

204. *Rule.* Multiply the rectangle of the segments into which the strained point divides the beam, in feet, by the weight in pounds, and divide this product by 139 times[47] the length in feet multiplied by the breadth in inches; the square root of the quotient will be the depth required in inches.

The depth between the upper and lower side will be $\cdot 7 \times$ by the whole depth. Consult art. 41 and 43 respecting the form, &c. of beams of this kind.

205. *Example.* Let C B, fig. 12, be 10 feet, and D C, 6 feet; and therefore B D the length, 16 feet; and the weight to be supported at A, 20,000 lbs., the breadth of the beam being 2 inches; required the depth?

By the rule

$$\frac{10 \times 6 \times 20000}{139 \times 16 \times 2} = 270;$$

and the square root of 270 is $16\frac{1}{2}$ inches nearly.

Also,

$$\cdot 7 \times 16\cdot 5 = 11\cdot 55 \text{ inches}$$

= the depth from a to b in fig. 12.

206 *Case* 3. When a load is distributed uniformly over the length of a beam. When the length is in feet, $p = \cdot 7$, and $f = 15{,}300$ lbs.,

---

[47] In practice it will be sufficiently accurate to use 140 for a divisor.

$$\frac{\mathrm{W}\,l}{1116} = b\,d^2,$$

by art. 117 and 139.

207. *Rule.* Multiply the whole weight in pounds by the length in feet; divide this product by 1116 times the breadth in inches, and the square root of the quotient will be the depth in inches.

Multiply this depth by ·7, which will give the depth between the upper and lower parts. Respecting the form of the beam, see art. 41.

208. *Example.* It is required to support a wall, 20 feet in height, and 18 inches in thickness, over an opening 26 feet wide, by means of two beams of cast iron, each 3 inches in thickness; required the depth?

Suppose a cubic foot of brick-work to weigh 100 lbs.; then

$$20 \times 1{\cdot}5 \times 26 \times 100 = 78000 \text{ lbs.}$$

the weight of the wall.

Therefore by the rule

$$\frac{78000 \times 26}{1116 \times 6} = 303 \text{ nearly};$$

and the square root of 303 is $17\frac{1}{2}$ inches, the depth required.

The depth between the upper and lower parts is

$$7 \times 17{\cdot}5 = 12{\cdot}25 \text{ inches.}$$

209. *Case* 4. When a beam is fixed at one end, and the load is applied at the other. Also, when the load acts at one end of a beam supported on a

centre of motion. By art. 117 we have, when the length is in feet, $p = \cdot 7$, and $f = 15{,}300$ lbs.,

$$\frac{W\,l}{139} = b\,d^2.$$

210. *Rule.* Calculate by the rule to Case 1, art. 201, using 140 instead of 560 for a divisor.

If the weight be uniformly distributed over the length of a beam fixed at one end, divide the weight by 2, and proceed as above directed.

DEFLEXION FROM CROSS STRAINS.

211. *Prop.* VII. To determine a rule for finding the deflexion of a cast iron beam, when the section is rectangular, and uniform throughout the length; the strain being 15,300 lbs. upon a square inch.

The same rules will apply to solid and hollow cylinders, to beams formed as figs. 9, 11, 12, and 26, when they are uniform throughout their length, and the depth used as a divisor is the greatest depth.

212. *Case* 1. When a beam is supported at the ends, and loaded in the middle, as in fig. 1.

By art. 121,

$$\frac{2\,\epsilon\,l^2}{3\,d} = \text{the deflexion,}$$

when $l =$ half the length; therefore,

$$\frac{3\,d \times D\,A}{2\,l^2} = \epsilon =$$

the greatest extension of an inch in length while the elastic force remains perfect. According to the experiment described in art. 56, the elastic force was perfect when the bar was loaded with 300 lbs.; hence we have

$$\frac{3\,d \times \mathrm{D\,A}}{2\,l^2} = \frac{3 \times 1 \times \cdot 16}{2 \times 17^2} = \frac{1}{1204} = \cdot 00083 \text{ inch}$$

$= \epsilon$ the extension of an inch in length, by a force equal to 15,300 lbs. upon a square inch; or generally, cast iron is extended $\frac{1}{1204}$ part of its length by a force equal to 15,300 lbs. upon a square inch.

If this value of ϵ be substituted in the equation, and l be made the whole length in feet, we have

$$\frac{2 \times \cdot 00083 \times 12^2 \times l^2}{3 \times 4 \times d} = \mathrm{D\,A}, \text{ or}$$

$$\frac{\cdot 01992\, l^2}{d} = \mathrm{D\,A};$$

hence it appears that the equation

$$\frac{\cdot 02\, l^2}{d} = \mathrm{D\,A}$$

may be used without sensible error.

Consequently, the deflexion of an uniform rectangular beam supported at the ends may be determined by the following rule:

213. *Rule.* Multiply the square of the length in feet by ·02; and this product divided by the depth in inches is equal to the deflexion in inches.

214. *Example.* Required the deflexion in the middle of a beam 20 feet long, and 15 inches deep, when strained to the extent of its elastic force?

By the rule

$$\frac{\cdot 02 \times 20^2}{15} = \cdot 533 \text{ inch};$$

therefore a beam loaded as in example (art. 147), will bend more than half an inch in the middle. If it be wished to reduce it to a quarter of an inch, double the breadth.

The deflexion of an uniform beam may also be found by Table II. art. 6.

215. *Case* 2. When an uniform rectangular beam is supported at the ends, and the load is equally distributed over the length. It has been shown in art. 139, Equation viii., that in this case the strain at any point is as the rectangle of the segments into which that point divides the beam; and the deflexion for that case is calculated by art. 126, Equation vii. And by comparing Equation ii. and vii.

$$\frac{2}{3} : \frac{5}{6} :: \frac{\cdot 02\, l^2}{d} : \frac{\cdot 025\, l^2}{d}.$$

Therefore the deflexion D A in the middle of a beam uniformly loaded is $= \frac{\cdot 025\, l^2}{d}$.

216. *Rule.* Multiply the square of the length in feet by ·025; and the quotient, from dividing this product by the depth in inches, will be the deflexion in the middle in inches.

217. *Example.* Let it be required to determine the deflexion that may be expected to take place in the example to Case 3, Prop. I. art. 152, where the length is 15 feet and the depth $12\frac{1}{4}$ inches?

By the rule

$$\frac{15 \times 15 \times \cdot 025}{12 \cdot 25} = \cdot 46 \text{ inch,}$$

the deflexion required.

218. This mode of calculation may often remove groundless alarm, as well as inform us when a structure is dangerous; for if a beam be loaded so as to bend more than is determined by the rule which applies to it, the structure may be justly deemed insecure. We also, by this mode of calculation, have an easy method of trying the goodness of a beam: for let it be loaded with any part, as for example $\frac{1}{4}$th of the weight it should bear, then the deflexion ought to be $\frac{1}{4}$th of the calculated deflexion. When a beam is tried by loading it with more than the weight it is intended to bear, it may be so strained as to break with the lesser weight, besides the difficulty and danger in trying such an experiment.

219. *Case* 3. When a beam is fixed upon a centre of motion, and the force applied at the other end, the flexure of the fixed part being insensible. The cranks of engines are in this case.

The flexure will be the same as in Case 1, art. 212, but the length of the beam being only half the length in that case, we have

$$\frac{\cdot 08\, l^2}{d} = \text{D A the deflexion.}$$

220. *Case* 4. If an uniform rectangular beam be fixed at one end, and the force be applied at the other, the deflexion of the end where the force is applied will be

$$\frac{\cdot 08\, l^2}{d} \times (1 + r).$$

For the deflexion from the extension of the projecting part of the beam is $\frac{\cdot 08\, l^2}{d}$, where l is the length of that part in feet; and if r be equal the $\frac{\text{length of fixed part,}}{l}$ then, by Equation iii. art. 133,

$$\frac{\cdot 08\, l^2}{d} \times (1 + r) = \text{the deflexion.}$$

221. *Rule.* Divide the length of the fixed part of the beam by the length of the part which yields to the force, and add 1 to the quotient; then multiply the square of the length in feet by the quotient so increased, and also by ·08; this product divided by the depth in inches will give the deflexion in inches.

222. *Example.* Conceive a beam, AB, fig. 26, to be uniform, and to be the beam of a pumping engine, the end B working the pumps, and the end A where the power acts 10 feet from the centre of motion, the end B 7 feet from the centre of motion, and the strain at B equal to the elastic force of the beam; through how much space will the point A move before the beam transmits the whole power to B, the depth of the beam being 12 inches?

In this case,

$$\frac{7}{10} = \cdot 7, \text{ and } 1 + \cdot 7 = 1\cdot 7;$$

therefore,

$$\frac{1\cdot 7 \times 10 \times 10 \times \cdot 08}{12} = 1\cdot 33 \text{ inches.}$$

223. *Prop.* VIII. To determine a rule for finding

the deflexion of a cast iron beam, of uniform breadth, when the outline of the depth is a parabola, the strain being equal to 15,300 lbs. per square inch.

The same rules will apply to beams of the form of section shown in figs. 9 and 11, when the breadth is uniform.

224. *Case* 1. When a beam is supported at the ends, and the load is applied in the middle.

The deflexion for this case is calculated in art. 123, Equation iv.; and comparing it with the deflexion of an uniform beam we have

$$\frac{2}{3} : \frac{4}{3} :: \frac{\cdot 02\, l^2}{d} : \frac{\cdot 04\, l^2}{d} = \text{the deflexion.}$$

225. *Rule*. Multiply the square of the whole length of the beam in feet by ·04; divide the product by the middle depth in inches, and the quotient will be the deflexion in inches.

226. *Example*. Let the depth of a beam be 18·4 inches, and its length 20 feet, which is on the supposition that the beam, of which the depth is found by example to Case 1, Prop. v. art. 188, is parabolic. By the rule

$$\frac{20 \times 20 \times \cdot 04}{18\cdot 4} = \cdot 87 \text{ inch,}$$

the deflexion required.

If the beam were of uniform depth, the deflexion would be only half this quantity, or ·435.

227. *Case* 2. If a parabolic beam of uniform breadth be fixed at one end, and the force be ap-

plied at the other, the deflexion of the end where the force is applied will be

$$\frac{\cdot 16\, l^2}{d}(1 + r),$$

where l is the length of the part the force acts on in feet, and $r =$ the quotient arising from dividing the length of the fixed part by the length l.

228. *Rule.* Divide the length in feet of the fixed part of the beam by the length in feet of the part which yields to the force, and add 1 to the quotient. Then multiply the square of the length in feet by the quotient so increased, and also by ·16; divide this product by the middle depth in inches, and the quotient will be the deflexion in inches.

229. *Example.* Let A B, fig. 26, be the beam of a steam engine, the moving force acting at A, and the resistance at B, C being the centre of motion; when A C = 12 feet, and C B = 10, and the depth in the middle 30 inches; it is required to determine the space the point A bends through before the full action is exerted on B, the strain being equal to the elastic force of the material?

In this case the length of the part C B, which may be considered as fixed, is 10 feet, and

$$\frac{10}{12} = \cdot 833, \text{ and } 1 + \cdot 833 = 1\cdot 833;$$

therefore,

$$\frac{12 \times 12 \times 1\cdot 833 \times \cdot 16}{30} = \frac{12 \times 22 \times \cdot 16}{30} = 1\cdot 408 \text{ inches},$$

the deflexion of the point A.

Few people are aware of the extent of flexure in

the parts of engines, and particularly when they are executed in a material which has been considered as nearly inflexible. In a well contrived machine, the importance of making the parts capable of transmitting motion and power with precision and regularity must be so obvious, that it appears almost incredible how much the laws of resistance have been neglected.

230. *Prop.* IX. To determine a rule for finding the deflexion of a cast iron beam of uniform breadth, when the depth at the end is only half the depth at the middle, the strain being equal to 15,300 lbs. on a square inch.

231. *Case* 1. When a beam is supported at the ends, and the load is applied in the middle. By art. 127, Equation viii.,

$$\frac{1{\cdot}09\, \epsilon\, l^2}{d} = \text{D A the deflexion;}$$

when this is compared with Equation ii. art. 121, we have

$$\frac{2}{3} : 1{\cdot}09 :: \frac{{\cdot}02\, l^2}{d} : \frac{{\cdot}0327\, l^2}{d} = \text{D A the deflexion.}$$

232. *Rule.* Multiply the square of the length in feet by ·0327, and the product divided by the depth in the middle in inches will give the deflexion in inches.

233. *Case* 2. When a beam is fixed at one end, and the force is applied at the other. In this case

$$\frac{{\cdot}13\, l^2}{d}(1 + r) = \text{the deflexion.}$$

234. *Rule.* Calculate the deflexion by the rule, art. 228, except changing the multiplier to ·13 instead of ·16.

235. *Prop.* x. To determine a rule for finding the deflexion of a beam, generated by the revolution of a cubic parabola, the strain being equal to 15,300 lbs. on a square inch.

The same rules will apply to any cases where the sections are similar figures, and the cube of the depth every where proportional to the leverage the force acts with.

236. *Case* 1. When a beam is supported at the ends, and the load is applied in the middle.

By art. 124, Equation v.,

$$\frac{6\,\epsilon\,l^2}{5\,d} = \text{D A the deflexion};$$

and comparing this with Equation ii. we have

$$\frac{2}{3} : \frac{6}{5} :: \frac{\cdot 02\,l^2}{d} : \frac{\cdot 036\,l^2}{d} = \text{the deflexion.}$$

237. *Rule.* Substitute ·036 in the place of ·04 in the rule to Prop. VIII. art. 225, and then calculate the deflexion by that rule.

238. *Case* 2. When a beam is fixed at one end, and the force acts at the other.

In this case

$$\frac{\cdot 144\,l^2}{d} = \text{the deflexion.}$$

239. *Rule.* In the rule to Prop. VIII. art. 228, use ·144 instead of ·16 as a multiplier, and calculate the deflexion by that rule, so altered.

240. *Prop.* XI. To determine a rule for finding the deflexion of a cast iron beam, of uniform breadth, the depth being bounded by an ellipse; the strain being equal to 15,300 lbs. on a square inch.

If the Equations ii. and vi. be compared, it will be found that

$$\frac{2}{3} : \cdot 857 :: \frac{\cdot 02\, l^2}{d} : \frac{\cdot 0257\, l^2}{d} = \text{the deflexion.}$$

241. *Rule.* The deflexion may be calculated by the rule to Prop. VIII. art. 225, if the multiplier ·0257 be employed instead of ·04.

242. *Prop.* XII. To determine a rule for the deflexion of a beam of uniform depth, when the breadth is bounded by a triangle, the strain upon a square inch being 15,300 lbs.

From Equations ii. and iii. art. 121 and 122, we have

$$\frac{2}{3} : 1 :: \frac{\cdot 02\, l^2}{d} : \frac{\cdot 03\, l^2}{d} = \text{the deflexion.}$$

243. *Case* 1. When a beam is supported at the ends, and loaded in the middle.

244. *Rule.* Calculate by the rule to Prop. VIII. art. 225, using ·03 instead of ·04 as a multiplier.

245. *Case* 2. When a beam is supported at one end, and fixed at the other.

In this case

$$\frac{\cdot 12\, l^2}{d} = \text{the deflexion.}$$

246. *Rule.* Calculate the deflexion by the rule to

Prop. VIII. art. 228, using 12 as a multiplier instead of ·16.

247. The rules derived from the twelve preceding propositions are applicable to any kind of material. For example, let it be required to adapt any one of the rules for oak: in the Alphabetical Table at the end of this Essay, art. OAK, it appears that oak is 0·25 as strong as cast iron; therefore, in a rule for strength, multiply the constant number by 0·25. Thus in the rules to Prop. I. Case 1,

$$850 \times 0{\cdot}25 = 212{\cdot}5,$$

the number to be used in these rules when the material is oak.

Again, oak is 2·8 times as extensible as cast iron; consequently the deflexion being found for cast iron, 2·8 times that deflexion will be the deflexion of oak, when it is strained to the extent of its elastic power.

SECTION VIII.

OF LATERAL STIFFNESS.

248. *Definitions.* The *stiffness* of a body is its resistance at a given deflexion. And the *lateral stiffness* is the stiffness to resist cross pressure.

249. *Prop.* XIII. *To determine the stiffness of an uniform bar or beam, of which the section is a rectangle, when fixed at one end, to resist a weight at the other; or supported in the middle on a centre to support a stress at each end.*

When a beam is strained to the extent of its elastic force, we have the weight it will bear, or

$$W = \frac{f\,b\,d^2}{6\,l},$$

(by art. 110,) and the deflexion under that strain will be

$$\frac{2\,\epsilon\,l^2}{3\,d} \times (1 + r),$$

(by art. 121 and 133, Equation iii.) Then, since the deflexion is proportional to the strain, if a be the given deflexion, and w the weight which produces it, we have

$$(1+r)\frac{2\,\epsilon\,l^2}{3\,d} : a :: \frac{f\,b\,d^2}{6\,l} : w = \frac{f\,b\,d^3\,a}{4\,\epsilon\,l^3\,(1+r)};$$

and because $\frac{f}{\epsilon} = m$, (art. 105,) we have

$$\frac{4\,w\,l^3\,(1+r)}{a\,m} = b\,d^3. \qquad \text{(i.)}$$

If the length be in feet, then

$$\frac{6912\,w\,\mathrm{L}^3\,(1+r)}{a\,m} = b\,d^3. \qquad \text{(ii.)}$$

Now in cast iron $m = 18{,}400{,}000$ ℔s.; therefore

$$\frac{w\,\mathrm{L}^3\,(1+r)}{2662\,a} = b\,d^3. \qquad \text{(iii.)}$$

Where $\mathrm{L} =$ the length in feet, a the deflexion in inches, b and d the breadth and depth in inches, and w the weight in pounds; and $r =$ the length of the fixed part divided by L. When $r = 1$, the lengths are equal, and $(1 + r) = 2$.

250. If the fixed part be of considerable bulk in respect to the other, we may neglect its effect on the deflexion, and in that case

$$\frac{w\,\mathrm{L}^3}{2662\,a} = b\,d^3. \qquad \text{(iv.)}$$

If in any of the preceding equations the breadth be diminished while the depth is uniform, the flexure will be increased; and when the outline of the breadths becomes a triangle, this increase is half the deflexion of a beam of uniform breadth; or the deflexions with the same strain are, as 2 : 3 (art. 122).

If the breadth be every where the same, but the beam be made the parabolic one of equal strength, then the deflexion will be twice as great as that of a beam of uniform depth (art. 123), and the general Equation iii. becomes

$$\frac{w\,L^3\,(1+r)}{1331\,a} = b\,d^3. \qquad \text{(v.)}$$

If the breadth be every where the same, but the outlines of the depth be straight lines, and the depth at either of the extremities half the depth at the point of greatest strain, then the deflexion is to that of a beam of uniform depth as 1·635 : 1 (art. 127); and Equation iii. becomes

$$\frac{w\,L^3\,(1+r)}{1628\,a} = b\,d^3. \qquad \text{(vi.)}$$

I shall illustrate this proposition by examples of its application to beams of pumping engines, cranks, and wheels.

BEAMS FOR PUMPING ENGINES.

251. *Example.* Let it be required to determine the breadth and depth of a beam for a pumping engine, its whole length being 24 feet, and the parts on each side of the centre of motion equal; and the straining force 30,000 lbs., the deflexion not to exceed 0·25 inch.

First, on the supposition that the beam is to be uniform, then, by Equation iii. art. 249,

$$\frac{w\,L^3\,(1+r)}{2662\,a} = \frac{30000 \times 12^3 \times (1+1)}{2662 \times \cdot 25} = b\,d^3 = 155790.$$

If the breadth be made 5 inches, the depth should be 31·5 inches; for

$$31{\cdot}5^3 \times 5 = 156279,$$

which very little exceeds 155,790.

But if the depth at the middle be double the depth at either end, use 1628 as a divisor instead of 2662; and calculating by Equation vi. we find $b\,d^3 = 254{,}742$, and if the breadth be 5 inches, the depth should be 37 inches.

CRANKS.

252. *Example.* If the force acting upon a crank be 6000 lbs., and its length be 3 feet, to determine its breadth and depth so that the deflexion may not exceed $\frac{1}{10}$th of an inch.

To this case, Equation iv. art. 250, applies, and

$$\frac{W\,L^3}{2662\,a} = \frac{6000 \times 3^3}{2662 \times {\cdot}1} = 653 = b\,d^3.$$

If the breadth be made 3 inches, the depth should be 6 inches, for the cube of $6 \times 3 = 648$.

When the depth at the end where the force acts is half the depth at the axis, use 1628 instead of 2662 for a divisor.

WHEELS.

253. For wheels, if N be the number of arms, or radii, our equation should be

$$\frac{W\,L^3}{2662\,N\,a} = b\,d^3$$

254. *Example* 1. Let the greatest force acting at the circumference of a spur-wheel be 1600 ℔s., the radius of the wheel 6 feet, and the number of arms 8; and let the deflexion not exceed $\frac{1}{10}$th of an inch.

Then by the Equation, art. 253,

$$\frac{W L^3}{2662 N a} = \frac{1600 \times 6^3}{2662 \times 8 \times \cdot 1} = b\,d^3 = 163.$$

If we make the breadth 2·5 inches, then

$$\frac{163}{2 \cdot 5} = 65 \cdot 2 = d^3;$$

and the cube root of 65·2 = 4·03 inches, nearly, for the depth or dimension of each arm, in the direction of the force.

When the depth at the rim is intended to be half that at the axis, use 1628 as a divisor instead of 2662 for a divisor.

If a wheel be strained till the arms break, the fractures take place close to the axis; there is a sensible strain at the part of the arm near the rim, but it is so small in respect to that at the axis, that its effect is neglected in our rule.

Example 2. When the stress on the teeth is 1090 ℔s. Suppose the wheel to be 4 feet radius, with 6 arms; and that a flexure of $\frac{2}{10}$ths of an inch will not sensibly affect the action of the wheel-work. Also, let the arms be diminished in depth so as to be only half the depth at the rim of the wheel; the breadth being fixed at 2 inches.

By the Equation, art. 253, we have

$$\frac{W L^3}{1628\, b\, a\, N} = \frac{1090 \times 64}{2 \times 1628 \times 6 \times \cdot 2} = 18 \text{ nearly} = d^3.$$

But the cube root of 18 is 2·62 inches; consequently next the axis the arms should be 2 by 2·62, and at the rim 2 by 1·31, in order that the play in applying the power may not exceed $\frac{2}{10}$ths of an inch. This rule gives the quantity of iron, with the rectangular section, but let it be disposed in the form of greatest strength consistent with that required for casting.

Again, let the pinion to be moved by the preceding wheel have a radius of 0·75 foot, with four arms, and the breadth of the arm 2 inches; the angular motion produced by the flexure being the same as above; that is, if

$$4 : \cdot 2 : : \cdot 75 : \cdot 0375 = a.$$

Then,

$$\frac{W L^3}{1628\, b\, a\, N} = \frac{4\, W L^2}{1628\, b\, N \times \cdot 2}; \text{ or,}$$

$$\frac{4 \times 1090 \times \cdot 56}{1628 \times 2 \times 4 \times \cdot 2} = \cdot 94 = d^3.$$

The cube root of ·94 is ·98 nearly, for the thickness of the arm at the axis.

255. I think we may in most cases allow a flexure of $\frac{2}{10}$ths of an inch for a wheel of 4 feet radius for the effect of the arms, and other $\frac{2}{10}$ths for the flexure of the shafts. In consequence, therefore, of such an arrangement, the strength of the arms will be expressed by a more simple equation; as well as the strength of the shafts to be treated in the section on Torsion.

When the flexure is 0·2 for a radius of 4 feet, it is very nearly a quarter of a degree; and with this degree of flexure, the arms of equal breadth, and the depth at the rim half the depth at the axis, we have

$$\frac{W L^2}{81 N} = b d^3. \quad \text{(vii.)}$$

Hence we have this practical rule. Multiply the stress at the pitch line in ℔s. by the square of the radius in feet; and divide the product by 81 times the breadth multiplied by the number of arms; and the cube root of the quotient will be the depth of the arm at the axis, and half this depth will be the depth at the rim.

If the thickness of the rim be made equal to the thickness of the teeth, and the breadth be proportioned by the Table, art. 158, then the number of arms should be $1\frac{1}{2}$ times the radius of the wheel in feet, divided by the square of the thickness of the teeth in inches, taken in the nearest whole numbers: it is usual to make an even number of arms; but there does not appear to be any reason for adhering to this practice. Wheels are often broke in the rim by wedging them on to the shaft; but the practice of fixing the wheels on by wedges has now given way to a much superior one, which consists in boring the eye truly cylindrical, and the shaft being turned to fit the eye, the wheel is retained in its place by a slip of iron, fitted into corresponding grooves in the shaft and in the eye of the wheel.

256. *Prop.* XIV. *To determine the stiffness of an uniform bar, or beam, supported at the ends, to resist a cross strain in the middle.*

If a beam be rectangular and uniform; then, making a the greatest deflexion that it ought to assume, we have by Equation ii. art. 121, and art. 137,

$$\frac{2\,\epsilon\, l^2}{12\, d} : a :: \frac{2 f b d^2}{3\, l} : w = \frac{4 f b d^3 a}{\epsilon\, l^3}.$$

And as

$$m = \frac{f}{\epsilon}, \text{(art. 105;)}\ w = \frac{4\, m b d^3 a}{l^3}.\text{[1]} \qquad \text{(viii.)}$$

When $L =$ the length in feet, then,

$$w = \frac{m\, b\, d^3\, a}{432\, L^3}. \qquad \text{(ix.)}$$

This equation answers for any material of which the weight of the modulus of elasticity is known; and this will be found in the Alphabetical Table at the end of this Work, for almost every kind of material in use. Its application to cast iron will be sufficient for an example.

---

[1] By some error of computation, Professor Leslie makes this equation

$$\frac{8\, m\, b\, d^3\, a}{5\, l^3} = w,$$

(Elements of Nat. Phil. vol. i. p. 237;) and, consequently, draws an erroneous measure of the modulus of elasticity from the experiments in art. 53 and 56. The equation I have arrived at is the same as Dr. Young had previously determined, (see his Nat. Philos. vol. ii. art. 326,) $2\,e$ in his equation being equal l in mine, and $2 f = w$.

The weight of the modulus for cast iron is 18,400,000 lbs., and, dividing this number by 432, we have for cast iron

$$w = \frac{42600\, b\, d^3\, a}{\mathrm{L}^3}. \qquad \text{(x.)}$$

257. If $a = \frac{\mathrm{L}}{40}$ of an inch, or the deflexion be as many fortieths of an inch as there are feet in the length of the beam; then the equation will be

$$w = \frac{1065\, b\, d^3}{\mathrm{L}^2};$$

which was made

$$\cdot 001\, w\, \mathrm{L}^2 = b\, d^3$$

to calculate the Table, art. 5. (xi.)

When the deflexion is only as many 100ths of an inch as the beam is feet in length, a deflexion which should not be greatly exceeded in shafts, on account of the irregular wear on their gudgeons and bearings when the flexure is greater, then

$$\frac{426\, b\, d^3}{\mathrm{L}^2} = w. \qquad \text{(xii.)}$$

If the load be uniformly distributed over the length of an uniform rectangular beam; then from art. 126 and 139, the dimensions being all in inches, we have

$$\frac{5\, \epsilon\, l^2}{24\, d} : a :: \frac{4 f\, b\, d^2}{3\, l} : w = \frac{32 f\, b\, d^3\, a}{5\, \epsilon\, l^3}.$$

And since

$$m = \frac{f}{\epsilon};\ w = \frac{32\, m\, b\, d^3\, a}{5\, l^3}. \qquad \text{(xiii.)}$$

Comparing this equation with Equation viii. it

appears that a weight uniformly distributed will produce the same depression in the middle as $\frac{5}{8}$ths of that weight applied in the middle, as has been otherwise shown by Dr. Young,[2] and Messrs. Barlow,[3] Dupin,[4] and Duleau.[5]

When w is the weight of the beam itself; then p being the weight of a bar of the same matter 12 inches long, and 1 square,

$$w = \frac{l\,b\,d\,p}{12};$$

and the deflexion of a beam by its own weight is

$$a = \frac{5\,\epsilon\,p\,l^4}{12 \times 32\,d^2 f} = \frac{5\,l^4}{384\,M\,d^2}. \qquad \text{(xiv.)}$$

Where M is the height of the modulus of elasticity[6] in feet (Equation v. art. 105).

258. In an uniform solid cylinder, the strength is to that of a square beam as 1 : 1·7 nearly (art. 113); therefore, by Equation x. art. 256, we have

$$\frac{w\,L^3}{25000\,a} = d^4. \qquad \text{(xv.)}$$

Where L is the length between the supports in feet, d the diameter in inches, and a the deflexion in inches produced by the weight w. in ℔s.

---

[2] Nat. Phil. vol. ii. art. 325 and 329.

[3] Treatise on the Strength of Timber, &c., art. 55. 1837.

[4] Idem, p. 97.

[5] Essai Théorique et Expérimental, art. 2 et 5.

[6] In theory this seems to furnish the most simple mode of obtaining the modulus, but it is not so accurate in practice, because it is difficult to ascertain the exact degree of flexure due to the weight.

If the load be uniformly diffused over the length, and s be the load on 1 foot in length in lbs.; then $w = \mathrm{L}\,s$, and the effect will be the same as if $\frac{5}{8}$ths of this load were applied in the middle, (art. 257;) consequently

$$\mathrm{L}\left(\frac{s}{40000\,a}\right)^{\frac{1}{4}} = d. \qquad \text{(xvi.)}$$

Therefore, if the load on a foot in length be the same, the diameter should be increased in direct proportion to the length, so that the flexure may be the same.

If in Equation xv. we make the flexure proportional to the length, and so that it may be $\frac{1}{100}$th of an inch for each foot in length.

Then,

$$\frac{w\,\mathrm{L}^2}{250} = d^4; \text{ or}$$

$$\sqrt{\mathrm{L}\left(\frac{w}{250}\right)^{\frac{1}{2}}} = d. \qquad \text{(xvii.)}$$

This equation will apply to uniform solid cylindrical shafts.

259. In a hollow shaft or cylinder, it will be only necessary to fix on what aliquot part of the diameter the thickness of metal should be, if its diameter were 1. Then, the difference between twice the thickness of metal and 1, will be the aliquot parts to be left hollow; and calling these parts n, it will be

$$\frac{d}{(1 - n^4)^{\frac{1}{4}}} =$$

the diameter of a hollow shaft of the same stiffness

as the solid one of the diameter *d*. (See Equation xviii. art. 115.) And the weight a solid shaft will sustain multiplied by $(1-n^4)$ will be the weight a hollow one of the same diameter will sustain.

Examples. If the thickness of metal be fixed at $\frac{1}{5}$th of the diameter, then

$$1-\frac{2}{5}=\frac{3}{5}=n=\cdot 6, \text{ and}$$

$$(1-\cdot 6^4)^{\frac{1}{4}}=\cdot 966=\frac{1}{1\cdot 0352}.$$

And if the diameter of a solid cylinder be found by Equation xv. xvi. or xvii., as the nature of the subject may require, and the diameter so found be multiplied by 1·0352, it will give the diameter of a hollow tube that will be of the same stiffness, the hollow part being $\frac{3}{5}$ths of the whole diameter.

In the same manner, if the thickness of metal be $\frac{1}{6}$th of the diameter, multiply by 1·056.

And if the thickness of metal be $\frac{3}{20}$ths of the diameter, multiply by 1·07.

The weight a hollow cylinder will sustain when the thickness of metal is exactly $\frac{1}{5}$th of the diameter, is 0·87 the weight a solid cylinder of the same external diameter would sustain with the same pressure; for $(1-n^4)=\cdot 87$. And its stiffness is to that of a square prism of the same depth as 1 is to 2, nearly.

260. *Example.* Required the diameter of a solid cylindrical shaft, 21 feet in length, that would not

be deflected more than half an inch by a weight of 31 cwt., or 3472 ℔s., applied in the middle.

By Equation xv. art. 258,

$$\frac{w\,L^3}{25000\,a} = \frac{3472 \times 21^3}{25000 \times \cdot 5} = d^4 = 2572, \text{ or } d = 7\cdot 12 \text{ inches,}$$

the diameter required.

261. *Example.* Required the diameter of a hollow shaft, 21 feet in length, the interior diameter $\frac{7}{10}$ths of the exterior one, that would not be deflected more than half an inch by a load of 3472 ℔s. applied in the middle of the length?

Find the diameter of the solid cylinder, as in the preceding example, and multiply it by 1·07 (see art. 259). That is,

$$7\cdot 12 \times 1\cdot 07 = 7\cdot 62 \text{ inches,}$$

the diameter required; the thickness of metal will be $\frac{3}{20}$ths of the diameter.

SECTION IX.

RESISTANCE TO TORSION.

262. *Definition.* The resistance which a shaft or axis offers to a force applied to twist it round is called the resistance to *Torsion.*

263. If a rectangular plate be supported at the corners A and B, fig. 29, Plate IV., and a weight be suspended from each of the other corners C D, then the strains produced by loading it in this manner will be similar to the twisting strain which occurs in shafts, and the like. In a cast iron plate the fractures would take place in the directions A B and C D at the same time; but, before the fracture, the one of the strains would serve as a fulcrum for the other; and the resistance to the forces at C and D would be sensibly the same as if the plate were supported upon a continued fulcrum in the direction A B.

Hence the strain may be considered a cross strain of the same nature as has been investigated in art. 108, and d D or c C the leverage the force at D or

C acts with, the breadth of the strained section being A B.

To find the breadth of the section of fracture, and the leverage in terms of the length and breadth of the plate, we have A B, the breadth, and by similar triangles,

$$\frac{A\,D \times B\,D}{A\,B} = D\,d \text{ the leverage.}$$

These values of the leverage and breadth being substituted in the Equation, art. 110, it becomes

$$W = \frac{f\,b\,d^2}{6\,l} = \frac{f\,d^2 \times A\,B \times A\,B}{6 \times A\,D \times B\,D};$$

or because

$$A\,B^2 = B\,D^2 + A\,D^2,$$

we have

$$W = \frac{f\,d^2}{6} \times \frac{B\,D^2 + A\,D^2}{A\,D \times B\,D}.$$

264. But when a force acts upon a shaft, it is usually at the circumference of a wheel upon that shaft, and if R be the radius of the wheel, then

$$\frac{2\,R\,W}{B\,D} =$$

the force collected at the surface of the shaft; and therefore, substituting this in the place of W, in the Equation above, we have

$$\frac{2\,R\,W}{B\,D} = \frac{f\,d^2}{6} \times \frac{B\,D^2 + A\,D^2}{A\,D \times B\,D};$$

$$\text{or, } W = \frac{f\,d^2}{12\,R} \times \frac{B\,D^2 + A\,D^2}{A\,D}.$$

If the length A D be l feet, and the leverage R

be in feet; then for cast iron $f = 15{,}300$ lbs., and we have

$$\frac{8{\cdot}85\, d^2\, (b^2 + 144\, l^2)}{R\, l} = W. \qquad \text{(i.)}$$

But this equation has a *minimum* value when $l = \frac{b}{12}$; therefore the resistance will be the same whatever the length may be, provided the length be not less than the breadth. Consequently, whenever the length exceeds the breadth, we have

$$\frac{212{\cdot}4\, d^2\, b}{R} = W.[1] \qquad \text{(ii.)}$$

But when b is to d in a less ratio than $\sqrt{2} : 1$ the shaft will not bear so great a strain, and it will bear least when its section is exactly square.

265. When a shaft is square, and its length l in feet, its side d in inches, and the leverage R in feet, then, from equation, art. 112, we obtain

$$W = \frac{f\, d^2}{3456\, R\, l} \times (2\, d^2 + 144\, l^2).$$

And when $f = 15{,}300$ lbs.,

$$W = \frac{8{\cdot}85\, d^2}{R\, l} \times (d^2 + 72\, l^2). \qquad \text{(iii.)}$$

In a square shaft also the resistance has a minimum value; that is, when $\sqrt{72}\, l = d$; hence, whenever the length is greater than the diagonal of the section, the strength will be

---

[1] In malleable iron the equation will be

$$\frac{238\, d^2\, b}{R} = W; \text{ for } 212{\cdot}4 \times 1{\cdot}12 = 238.$$

$$\frac{150\,d^3}{R} = W.^{2} \qquad \text{(iv.)}$$

Where R is the radius of the wheel in feet to which the power W in pounds is to be applied, and d is the side of the shaft or axis in inches.

266. In a cylindrical shaft the section of fracture is an ellipse, and when l and R are in feet, and $f = 15{,}300$, d being the diameter of the shaft in inches, we have by art. 114,

$$W = \frac{5{\cdot}2\,d^2}{R\,l} \times (d^2 + 144\,l^2). \qquad \text{(v.)}$$

267. Here again it may be shown, by the principles of maxima and minima, that there is a particular line of fracture where the resistance to torsion is a minimum; in a cylindrical body this happens when $12\,l = d$; that is, when the length is equal to the diameter.

Consequently, in all cases where the length exceeds the diameter, the equation in art. 266 should be applied in the form

$$\frac{124{\cdot}8\,d^3}{R} = W.^{3} \qquad \text{(vi.)}$$

As the equation reduces to this form by substituting $\frac{d}{12}$ for l.

---

[2] In malleable iron shafts the equation will be

$$\frac{168\,d^3}{R} = W.$$

[3] For malleable iron make $\frac{140\,d^3}{R} = W.$

268. In the same manner it may be shown that in a tube or hollow cylinder of which the length is greater than the diameter, the resistance to torsion is expressed by the equation

$$\frac{124{\cdot}8\, d^3 (1 - n^4)}{R} = W. \qquad \text{(vii.)}$$

Where d is the exterior diameter in inches, and $n\, d$ the interior diameter.

It will be a good proportion in practice to make $n = 0{\cdot}6$; then the rule becomes

$$\frac{108\, d^3}{R} = W. \qquad \text{(viii.)}$$

Where d is the exterior diameter in inches, and the thickness of metal is exactly $\frac{1}{5}$th of the diameter; R, as before, being the radius of the wheel in feet, to the circumference of which the power W in lbs. is applied.

269. *Example.* Let it be required to find the diameter of a shaft for a water-wheel, the radius of the water-wheel 9 feet, and the greatest force that it will be exposed to at the circumference, 2000 lbs.

If the shaft is to be a solid cylinder, then the diameter will be found by Equation vi. art. 267; that is,

$$\frac{W\,R}{124{\cdot}8} = \frac{2000 \times 9}{124{\cdot}8} = 144{\cdot}2 = d^3.$$

And the cube root of 144·2 is $5\frac{1}{4}$ inches, the diameter required.

If the shaft is to be a hollow cylinder, Equation viii. will apply, where

$$\frac{W\,R}{108} = \frac{2000 \times 9}{108} = 166 \cdot 7 = d^3.$$

And the cube root of 166·7 is $5\frac{1}{2}$ inches the diameter, when the thickness of metal is $\frac{1}{5}$th of this diameter.

270. But the lateral stress on a shaft will always be greater than the twisting force, when the length of the shaft exceeds $\frac{1}{4}$th of the radius of the wheel; yet the preceding equations will often be of use in calculating the strength of journals,[4] and these calculations should be made by Equation vi. in the same manner as in the example in the preceding article; only as an allowance for wear the diameter should be $\frac{1}{6}$th greater than is given by the rules.

271. The preceding investigation has been confined to the strength to resist twisting, but in shafts of great length in respect to their diameters, the effect of flexure is considerable.

Let ϵ be the extension the material will bear without injury when the length is unity. This extension must obviously limit the movement of torsion, or the angle of torsion. But, since the line of greatest strain, in a bar of greater length than its diameter, is always in the direction of the diagonal of a square; if a square were drawn on the surface of the bar in its natural state, it would become a rhombus by the action of the straining force, and the quantity of angular motion would be nearly $\sqrt{2}$ times

[4] A journal is different from a gudgeon only in being exposed to a considerable twisting strain.

the extension of the diagonal; or twice the extension of the length of the bar. For if a line were wound round the bar at an angle of 45° with the axis, its length would be $l\sqrt{2}$; l being the length of the bar in feet. Therefore, $l\,\epsilon\sqrt{2}=$ the extension, and $2\,l\,\epsilon$ the arc described in feet, or $24\,l\,\epsilon=$ the arc in inches. But if a be the number of degrees in an arc, and $\frac{d}{2}$ its radius; ·0174533 being the length of an arc of one degree when its radius is unity; we have

$$24\,l\,\epsilon = \frac{a\,d}{2} \times {\cdot}0174533; \text{ or}$$

$$\frac{2750\,l\,\epsilon}{d} = a. \qquad \text{(ix.)}$$

That is, the angle of torsion a is as the length and extensibility of the body directly, and inversely as the diameter.

If the value of ϵ be taken for cast iron, that is, $=\frac{1}{1204}$, we have

$$\frac{2{\cdot}284\,l}{d} = a.\text{[5]} \qquad \text{(x.)}$$

Here l is the length of the shaft or other body in feet; d its diameter in inches, and a the angle of torsion in degrees of a circle.

Example. Thus, let the vertical shaft of a mill be 30 feet in length, and the diameter 10 inches; then, when it is strained to the extent of its elastic force,

---

[5] In malleable iron, $\epsilon = \frac{1}{1400}$; therefore, $\frac{1{\cdot}965\,l}{d} = a$.

$$\frac{2{\cdot}284 \times 30}{10} = 6\tfrac{3}{4} \text{ degrees nearly.}$$

In certain cases this degree of twisting may be of considerable advantage in preventing the shocks incidental to machines moved by wind, horses, or other irregular powers; but in other cases it will be a disadvantage, because the motion will neither be so accurate, nor so certain to produce the desired effect.

272. Since the angle of torsion is as the extension, it will be as the straining force; and to estimate the stiffness of a body to resist torsion, we have this analogy when the body is a hollow cylinder; from Equation vii. and ix. of this section,

$$\frac{2750\, l\, \epsilon}{d} : a :: \frac{124{\cdot}8\, d^3\, (1-n^4)}{\mathrm{R}} : \mathrm{W} = \frac{124{\cdot}8\, d^4\, a\, (1-n^4)}{2750\, \mathrm{R}\, l\, \epsilon}.$$

Or more generally,

$$\frac{f\, d^4\, a\, (1 - n^4)}{336600\, \mathrm{R}\, l\, \epsilon} = \mathrm{W}.$$

And if m be the weight of the modulus of elasticity, (art. 105,)

$$\frac{m\, d^4\, a\, (1 - n^4)}{336600\, l\, \mathrm{R}} = \mathrm{W}. \qquad \text{(xi.)}$$

When $n = o$ the equation applies to a solid cylinder.

When a shaft is rectangular, the analogy from Equation ii. and ix. becomes

$$\frac{2750\, l\, \epsilon}{b} : a :: \frac{212{\cdot}4\, d^2\, b}{\mathrm{R}} : \mathrm{W} = \frac{212{\cdot}4\, d^2\, b^2\, a}{2750\, \mathrm{R}\, l\, \epsilon}; \text{ or}$$

$$\frac{m\, d^2\, b^2\, a}{198900\, \mathrm{R}\, l} = \mathrm{W}. \qquad \text{(xii.)}$$

We have now to show the application of these equations, and to form practical rules from them. The value of m for cast iron is 18,400,000 lbs.; consequently Equation xi. applied to cast iron is

$$\frac{55\ d^4\ a\ (1 - n^4)}{l\ \mathrm{R}} = \mathrm{W}.\text{[6]} \qquad \text{(xiii.)}$$

And Equation xii. gives

$$\frac{92{\cdot}5\ d^2\ b^2\ a}{l\ \mathrm{R}} = \mathrm{W}.\text{[7]} \qquad \text{(xiv.)}$$

PRACTICAL RULES AND EXAMPLES FOR THE STIFFNESS OF CYLINDRICAL SHAFTS TO RESIST TORSION.

273. In practical cases there will be known the length of the shaft, the power, and the leverage the power acts with; and there must be fixed, by the person who applies the rule, the number of degrees of torsion that will not affect the action of the machine; this being settled, the diameter of the shaft will be determined by the rule.

Rule 1. To determine the diameter of a solid cylinder to resist torsion, with a given flexure.

Multiply the power in pounds by the length of the shaft in feet, and by the leverage in feet. Divide this product by 55 times the number of degrees in the angle of torsion, which is considered best for

---

[6] In malleable iron,

$$\frac{74\ d^4\ a\ (1 - n^4)}{\mathrm{R}} = \mathrm{W}.$$

[7] In malleable iron,

$$\frac{124\ d^2\ b^2\ a}{l\ \mathrm{R}} = \mathrm{W}.$$

the action of the machine; and the fourth root of the quotient will be the diameter of the shaft.

Example. Let it be required to find the diameter for a series of lying shafts 30 feet in length to transmit a power equal to 4000 lbs. acting at the circumference of a wheel of 2 feet radius, so that the twist of the shafts on the application of the power may not exceed one degree?

Here the whole length must be taken as if it were one shaft, and by the rule,

$$\frac{4000 \times 30 \times 2}{55 \times 1} = 4364,$$

and by a Table of powers,[8] the fourth root is found to be 8·13 inches, the diameter required.

If the machinery be required to act with much precision, this will be as much flexure as can be allowed; but in ordinary cases two degrees might be admitted, and then a little less than 7 inches would be the diameter.

Where there is much wheel-work, the flexures should be less; indeed it does not appear to be desirable to exceed a quarter of a degree for the shafts or axes.

274. *Rule* 2. To determine the diameter of a hollow cylinder to resist torsion, when the thickness of metal is $\frac{1}{5}$th of the diameter, and the flexure given.

Multiply the power in pounds by the length of

---

[8] See Barlow's Mathematical Tables, Table III.

the shaft in feet, and by the leverage the power acts with in feet. Divide the product by 48 times the angle of flexure in degrees; the fourth root of the quotient will be the diameter required in inches.

Example. Let the diameter of a hollow shaft be determined, so that it may be sufficient to withstand a force of 800 lbs. acting at the circumference of a wheel of 4 feet radius with a flexure of one degree; the thickness of metal to be $\frac{1}{5}$th of the diameter, and the length 10 feet.

In this case

$$\frac{800 \times 10 \times 4}{48 \times 1} = 666{\cdot}6;$$

and the fourth root of 666·6 is 5·1 inches nearly; which is the diameter required.

PRACTICAL RULE AND EXAMPLE FOR THE STIFFNESS OF SQUARE SHAFTS TO RESIST TORSION.

275. Rules for square shafts are applications of Equation xiv.; and the same things are known as in the case of cylindrical shafts.

Rule. To determine the side of a square shaft to resist torsion with a given flexure.

Multiply the power in pounds by the leverage it acts with in feet, and also by the length of the shaft in feet. Divide this product by 92·5 times the angle of flexure in degrees, and the square root of the quotient will be the area of the shaft in inches.

Example. Suppose the length of a shaft is to be 12 feet, and it is to be driven by a power of 700 lbs.

acting on a pinion, on the shaft, of 1 foot radius to the pitch line, and that a flexure of 1 degree will not affect the machinery.

By the rule,

$$\frac{700 \times 1 \times 12}{92{\cdot}5 \times 1} = 90{\cdot}8.$$

The square root of 90·8 is 9·53, the area of the section in inches; and the square root of 9·53 is 3·1 inches nearly, for the side of the shaft.

The reader may find further information on the torsion of wires, and the laws of the oscillation of the torsion balance, in Dr. Young's Lectures on Nat. Philos. vol. i. pp. 140, 141; Dr. Brewster's Edinburgh Encyclopædia, art. Mechanics, p. 544 to 549; Dr. Brewster's edition of Ferguson's Lectures, vol. ii. p. 234; or Professor Leslie's Elements of Nat. Philos. vol. i. p. 243.

SECTION X.

OF THE STRENGTH OF COLUMNS, PILLARS, OR OTHER SUPPORTS COMPRESSED, OR EXTENDED, IN THE DIRECTION OF THEIR LENGTH.

276. If the length of a column be considerable with respect to its diameter, under a certain force it will bend; but when it becomes too short to bend, its strength is only limited by the force which would crush it. Considering, however, that it is imprudent to load even a short column beyond its elastic force, an inquiry respecting the phenomena of crushing would lead to nothing useful.

Let A A be a column, fig. 30, supported at A′, and supporting a load at A; and let this load have produced its full effect in straining the column. Let E be the neutral axis, B and D the centres of resistance, and A F the direction of the straining force. Draw d D parallel to A F, then, by the principles of statics, we have

$$d\,\mathrm{D} : \mathrm{D\,A} :: \mathrm{W}\ (\text{the weight}) : \frac{\mathrm{W.\,D\,A}}{d\,\mathrm{D}} =$$

the compressive force in the direction A D. Also,

$$\mathrm{DA : AF :: \frac{W.DA}{\mathit{d}D} : \frac{W.AF}{\mathit{d}D}} =$$

the vertical pressure at D.

But, by similar triangles,

$$\mathrm{BD : BF :: \mathit{d}D : AF} = \frac{\mathrm{BF.\mathit{d}D}}{\mathrm{BD}},$$

therefore

$$\frac{\mathrm{W.AF}}{\mathit{d}\mathrm{D}} = \frac{\mathrm{W.BF}}{\mathrm{BD}}. \qquad \text{(i.)}$$

277. In a similar manner it may be proved that the strain at B is expressed by

$$\frac{\mathrm{W.}\overline{\mathrm{BF - BD}}}{\mathrm{BD}}. \qquad \text{(ii.)}$$

Where it is obvious that when $\mathrm{BD = BF}$ this strain is nothing; that is, when the direction of the straining force passes through the point D, or the neutral axis coincides with the surface of the block. It also may be observed, that when B F exceeds B D, this strain is expressed by a positive quantity, indicating extension; but when B F is less than B D it is negative, indicating that it is a resistance to compression. If $\mathrm{BF} = \frac{1}{2}\,\mathrm{BD}$, then both points are equally compressed.

The force has been supposed to be perpendicular to the plane of section for which the strains have been calculated, but this is not essential to the investigation; it is only the most usual strain on columns and ties. For instead of the force acting in the direction A F, let the force act in the direction A G; and let the angle F A G be denoted by C. Then, the investigation being resumed, we shall find Equation i. will become

$$\frac{(\mathrm{B\,F} + \mathrm{A\,F.\ sin.\ C})\ \mathrm{W.\ cos.\ C}}{\mathrm{B\,D}} = \text{stress at D.} \qquad \text{(iii.)}$$

And,

$$\frac{(\mathrm{B\,F} + \mathrm{A\,F.\ sin.\ C} - \mathrm{B\,D})\ \mathrm{W.\ cos.\ C}}{\mathrm{B\,D}} = \text{stress at B.} \qquad \text{(iv.)}$$

But when the force acts in an oblique direction, a further stay is required to prevent the pillar overturning; and wherever this stay is placed there the greatest strain would be in a straight pillar. If it be stayed at D F, then the stay placed there becomes a point of support; and the action of the forces on the beam are similar to those already considered in fig. 14, Plate II. But it was remarked in a note to art. 108, that the mode of calculation there given is not correct when the beam is not nearly horizontal; the difference is owing to a change of the position of the neutral axis caused by the oblique direction of the force. The position of that axis, and the strength of the section, we will now proceed to calculate; and to develope the changes produced by altering the direction of the straining force.

278. It may be shown that the resistance of the section, on either side of the neutral axis, is equal to the force of a square inch multiplied by the area of that section and by the distance of the centre of gravity from the neutral axis, and divided by the

distance of the compressed surface from the neutral axis, when B or D is the centre of percussion of the section.[1]

279. Let x be the distance of the neutral axis from the middle of the depth; $y = EG$ the distance of the direction AG of the straining force from the middle of the depth; $d =$ the depth, $b =$ the breadth, and f the resistance of a square inch; then the area of the compressed part of the section will be $(\frac{1}{2}d + x)\,b$, and the extended part of the section $(\frac{1}{2}d - x)\,b$. Therefore, if $n\,(\frac{1}{2}d + x)$ and $n\,(\frac{1}{2}d - x)$ be the distances of the centres of percussion from the neutral axis, and $m\,(\frac{1}{2}d + x)$ and $m\,(\frac{1}{2}d - x)$ the distances of the centres of gravity, we shall have

$$\frac{W.\,BG\cos.C}{BD} \times \frac{mfb\,(\frac{1}{2}d - x)^2}{(\frac{1}{2}d + x)} = \frac{W.\,(BG - BD)\cos.C}{BD} \times \frac{mfb\,(\frac{1}{2}d + x)^2}{(\frac{1}{2}d + x)}, \text{ or}$$

$$BG \times (\tfrac{1}{2}d - x)^2 = (BG - BD) \times (\tfrac{1}{2}d + x)^2;$$

and when the proper substitutions are made, this equation reduces to

$$x^2\,(2 - 3n) + 2yx - \tfrac{1}{4}nd^2 = 0.$$

280. In a rectangular section $n = \frac{2}{3}$, and consequently we find by the preceding equation

[1] Emerson's Mechanics, 4to. edit. Prop. LXXVII.

$$x = \frac{d^2}{12\,y}.$$ [2]

Also, since in this case $m = \frac{1}{2}$, we have an equilibrium between the compressing force and the resistance to compression, when

$$\frac{\text{W. B G. cos. C}}{\text{B D}} = \frac{f\,b}{2}\left(\tfrac{1}{2}\,d + x\right);$$

and substituting for B G, B D, and x their proper values, this equation becomes

$$\text{W} = \frac{f\,b\,d^2}{(d + 6\,y)\text{ cos. C}}. \qquad \text{(v.)}$$

But if B F be denoted by a; and $\frac{l}{2} = \text{A F}$, whence F G will be =

$$\frac{\text{A F. sin. C}}{\text{cos. C}} = \frac{l\text{ sin. C}}{2\text{ cos. C}};$$

and therefore,

$$y = a + \frac{l\text{ sin. C}}{2\text{ cos. C}}.$$

---

[2] It is shown that

$$y = a + \frac{l\text{ sin. C}}{2\text{ cos. C}} = a + \tfrac{1}{2}\,l\text{ tan. C};$$

hence the distance of the neutral axis from the axis of the column is

$$x = \frac{d^2}{12\,(a + \frac{1}{2}\,l\,\text{tan. C})};$$

and these axes must coincide when C is an angle of 90°, that is, when the direction of the force is perpendicular to the axis of the column; but not in any other case.

For when C = 90°, the tan. C is unlimited; and consequently the fraction which represents x is incomparably small, or the axes coincide.

Consequently,

$$\frac{f\,b\,d^2}{(d+6\,y)\cos.\,C} = \frac{f\,b\,d^2}{d.\cos.\,C + 6\,a.\cos.\,C. + 3\,l\sin.\,C} = W. \quad \text{(vi.)}$$

This equation will enable us to trace the particular conditions of this important problem.

In the first place, if the points E and A be in a line perpendicular to B G, then $a = o$, and the equation is

$$\frac{f\,b\,d^2}{d.\cos.\,C + 3\,l\sin.\,C} = W. \quad \text{(vii.)}$$

Secondly, if the force act in a direction parallel to B G, then C = 90 degrees; and sin. C = 1, and cos. C = o, and Equation vi. becomes

$$\frac{f\,b\,d^2}{3} = W. \quad \text{(viii.)}$$

We have in this case the same equation as in art. 110, for in this instance l is double the length taken in that equation.

Thirdly, if the force act in a direction perpendicular to B G, then cos. C = 1, and sin. C = o, and consequently Equation vi. becomes

$$\frac{f\,b\,d^2}{d + 6\,a} = W. \quad \text{(ix.)}$$

Fourthly, when $a = o$, or the direction of the force coincides with the axis E, then

$$f\,b\,d = W. \quad \text{(x.)}$$

And, fifthly, if a = half the depth of the block, then

$$\frac{f\,b\,d}{4} = W. \quad \text{(xi.)}$$

The Equations ix. x. and xi. apply to short columns, or blocks, of which the length is not more than ten or twelve times the least dimension of the section; and from them are derived the following practical rules:

TO FIND THE AREA OF A SHORT RECTANGULAR COLUMN OR BLOCK TO RESIST A GIVEN PRESSURE.

281. *Rule.* When the force is to be applied exactly in the axis or centre of the section of the block, divide the pressure or the weight in pounds by 15,300, and the quotient will be the area of the section of the block in inches. But since this requires a degree of precision in adjusting the direction of the force which it is altogether impossible to arrive at in practice, and when a force presses a block of which $a\,a'$ is the axis, fig. 31, Plate IV., it is always probable that the direction A A′ of the force may act upon one edge only of the end of the block, and therefore be at a distance of half the least thickness from the axis; which will reduce the resistance of the block to $\frac{1}{4}$th, and consequently the area should always be made four times as great as is determined by this rule.

When the distance of the direction of the force from the axis is determined by the nature of the construction, the following is a general rule.

282. *Rule.* To the thickness (or least dimension of the section) in inches, add six times the distance

of the direction of the force from the axis in inches, and let this sum be multiplied by the weight or pressure in pounds; divide the quotient by 15,300 times the square of the least thickness in inches, and the quotient will be the breadth of the block in inches.

This rule is the Equation ix. art. 280, in words at length, and it applies to resistance to tension as well as to resistance to compression.

283. The writer of the article 'Bridge,' in the Supplement to the Encycl. Brit., has shown that when the force acts in the direction of the diagonal of the block, as is shown in fig. 32, the strain will be twice as great as when the same force acts in the direction of the axis.[3] Now the reader will be satisfied, that, in consequence of settlements, or other causes, a column is always liable to be strained in this manner; and therefore will carefully avoid enlarging the ends of his columns, under the notion of gaining stability, for the effect of the straining force will be still more increased by such enlargement in the event of a change of direction from settlement, as in fig. 33. In my 'Treatise on Carpentry,' I have recommended circular abutting joints to lessen the effect of a partial change in the position of the strained pieces,[4] an idea which appears to have occurred, in the first instance, to Serlio.[5]

---

[3] Napier's Supp. to Encycl. Brit., art. 'Bridge,' Prop. I. p. 499.

[4] Tredgold's Elementary Prin. Carpentry, Sect. IX. p. 164. 1840.

[5] Serlio's Architecture, Lib. I. p. 13. Paris, 1545.

284. A general solution of the equation expressing the stress and strain, when the column is cylindrical, is complicated, but in one particular case the result is extremely simple; that is, when the neutral axis is in one of the surfaces of the column. If d be the diameter of the column, then 7854 d^2 = the area, and $\frac{1}{2} d$ = the distance of the centre of gravity, and therefore

$$\frac{\text{W. B G. cos. C}}{\text{B D}} = \frac{\cdot 7854\, d^2 f}{2}.$$

But when the neutral axis is in the surface of the cylinder,

$$\text{B G} = \text{B D, or W} = \frac{\cdot 7854\, d^2 f}{2 \text{ cos. C}}.$$

In this case the distance of the direction of the force from the axis of the column will be $\frac{1}{8}$th of the diameter, the centre of percussion being $\frac{5}{8} d$ distant from the neutral axis.

285. Hence it appears, that when the distance of the direction of the force from the axis is $\frac{1}{8} d$, the strength of a cylinder is to that of a circumscribed square prism, as seven times the area of the cylinder, to eight times the area of the prism; or nearly as 5·5 : 8, or as 1 : 1·46 nearly.

When the neutral axes are at or near the axes of the pieces, the ratio of the strength of the cylinder to that of the prism becomes

$$\frac{3 \times \cdot 7854}{4} : 1, \text{ or as } 1 : 1\cdot 7.$$

as has been shown by Dr. T. Young;[6] consequently in a column, when both the resistances to compression and extension are brought into action, the ratio varies between 1 : 1·46 and 1 : 1·7; the mean being nearly 1 : 1·6.

OF THE STRENGTH OF LONG PILLARS AND COLUMNS.

286. If a support be compressed in the direction of its length, and the deflexion be sufficient to sensibly increase the distance of the direction of the force from the axis, in the middle of the length of the support, it is evident that the strain will be increased; and since the curvature in practical cases will be very small, we may suppose it to be an arc of a circle. In a circle the square of the length of the chord, in a small segment, is sensibly equal to the radius $\times$ 8 times the versed sine; or $\frac{l^2}{8\delta} = \text{radius}$. The deflexion will be greatest when the neutral axis coincides with the axis, and taking this extreme case, we shall have this analogy;—as the alteration of the length of the concave side is to the original length, so is the $\frac{1}{2}$ depth to the radius of curvature; or,

$$\epsilon : 1 :: \frac{d}{2} : \text{radius} = \frac{d}{2\epsilon}.$$

Therefore

$$\frac{l^2}{8\delta} = \frac{d}{2\epsilon}; \text{ and } \delta = \frac{l^2\epsilon}{4d} = \text{the deflexion in the middle.}$$

[6] Dr. Young's Lectures on Nat. Philos. vol. ii. art. 339, B.

287. Let the distance of the direction of the force from the axis, when first applied, be denoted by a, as in a preceding article, (art. 280;) then, in consequence of the flexure, it will be equal to

$$a + \frac{l^2 \epsilon}{4 d};$$

consequently by Equation ix. we have

$$\frac{f b d^2}{d + 6 a + \frac{6 l^2 \epsilon}{4 d}} = W. \qquad \text{(xii.)}$$

In cast iron $f = 15{,}300$ ℔s. and $\epsilon = \frac{1}{1204}$, (art. 143 and 212;) therefore, if l be the length in feet, b, d, and a in inches, we obtain the following practical formula, for the strength of a rectangular prism, viz.

$$\frac{15300 b d^2}{d + 6 a + \frac{18 l^2}{d}} = \frac{15300 b d^3}{d^2 + 6 d a + \cdot 18 l^2} = W.\text{[7]} \qquad \text{(xiii.)}$$

288. If $a = o$, or the direction of the force coincides with the axis, then the rule becomes

$$\frac{15300 b d^3}{d^2 + \cdot 18 l^2} = W. \qquad \text{(xiv.)}$$

It would, however, be improper in practice to calculate upon the nice adjustment of the direction of the pressure in the direction of the axis, which is

---

[7] For malleable iron,

$$\frac{17800 b d^3}{d^2 + 6 d a + \cdot 16 l^2} = W.$$

For oak,

$$\frac{3960 b d^3}{d^2 + 6 d a + \cdot 5 l^2} = W.$$

supposed in the preceding equation; indeed, there are very few instances where its direction may not in all probability be at the distance of half the depth from the axis, and in that case $a = \frac{1}{2}d$, and

$$\frac{15300\, b\, d^3}{4\, d^2 + \cdot 18\, l^2} = \text{W.}^{8} \qquad \text{(xv.)}$$

289. As an approximate rule for the strength of a cylinder to resist compression in the direction of its length, we have

$$\frac{15300\, d^4}{1\cdot 6\, (d^2 + \cdot 18\, l^2)} = \frac{9562\, d^4}{d^2 + \cdot 18\, l^2} = \text{W.} \qquad \text{(xvi.)}$$

290. And if the direction of the force be a inches distant from the axis, the rule is

$$\frac{9562\, d^4}{d^2 + 6\, d\, a + \cdot 18\, l^2} = \text{W.} \qquad \text{(xvii.)}$$

If the force act in the direction of one of the surfaces of the column, then $a = \frac{1}{2}d$, and

$$\frac{9562\, d^4}{4\, d^2 + \cdot 18\, l^2} = \text{W}^{9} \qquad \text{(xviii.)}$$

By this rule the Table of columns (Table III.

---

[8] In malleable iron,

$$\frac{17800\, b\, d^3}{4\, d^2 + \cdot 16\, l^2} = \text{W.}$$

In oak,

$$\frac{3960\, b\, d^3}{d^2 + \cdot 5\, l^2} = \text{W.}$$

[9] In malleable iron,

$$\frac{11125\, d^4}{4\, d^2 + \cdot 16\, l^2} = \text{W.}$$

In oak,

$$\frac{2470\, d^4}{d^2 + \cdot 5\, l^2} = \text{W.}$$

p. 26,) was calculated, only the weight is there given in cwts.

In all the rules from Equation xiii. to xviii. l is the length, A A′, fig. 31, Plate IV., in feet, d either the diameter or the least side in inches, b the greater side in inches, and W the weight to be supported in ℔s.

291. *Example* 1. Required the weight that could be supported, with safety, by a cylindrical column, the length being 11 feet, and the diameter 5 inches, and supposing it probable that the force may act in the direction A A′, fig. 31, at the distance of half the diameter from the axis?

In this example Equation xviii. art. 290, should be used; and therefore

$$\frac{9562\,d^4}{4\,d^2 + \cdot 18\,l^2} = \frac{9562 \times 5^4}{4 \times 5^2 + \cdot 18 \times 11^2} = \text{W} = 49{,}080 \text{ ℔s.}$$

or a little above 22 tons.

In this manner may be calculated the strength of story-posts for supporting buildings. When they are for houses, ample allowance should be made for the weight of crowded rooms, and when for warehouses the greatest possible weight of goods should be estimated.

292. *Example* 2. It is proposed to determine the compression a curved rib will sustain in the direction of its chord; the greatest distance of the axis of the rib from the chord line being 6 inches, the size of the rib 3 inches square, and the length of the chord line 5 feet.

By Equation xiii. art. 287,

$$W = \frac{15300\, b\, d^2}{d^2 + 6\, d\, a + \cdot 18\, l^2} = \frac{15300 \times 3^4}{3^2 + 6 \times 3 \times 6 + \cdot 18 \times 5^2} = 30{,}600 \text{ lbs.}$$

Example 3. The piston-rod of a double acting steam engine is another interesting case to which these equations will apply; and the reader will excuse my having recourse to algebraic notation in order to make the rule general.

Let D be the diameter of the steam cylinder in inches, and p the greatest pressure of the steam on a circular inch of the piston in lbs. Then $W = D^2 p$.

But it has been shown in a note to art. 290, that in malleable iron

$$W = \frac{11125\, d^4}{4\, d^2 + \cdot 16\, l^2}.$$

Therefore,

$$D^2 p = \frac{11125\, d^4}{4\, d^2 + \cdot 16\, l^2}; \text{ or}$$

$$D = 53\, d^2 \sqrt{\frac{1}{p\,(d^2 + \cdot 04\, l^2)}}.$$

Now in an extreme case we can never have the length in feet greater than about three times the diameter in inches; substitute this value of l, and we have

$$\frac{D \sqrt{1 \cdot 5\, p}}{53} = d.$$

If the pressure be 8 lbs. on the circular inch, that is, a little more than 10 lbs. on the square inch, it gives $\frac{D}{15} = d$. That is, the piston rod should never be less than $\frac{1}{15}$th of the diameter of the cylinder in a double acting steam engine. In practice it is usual to make them $\frac{1}{10}$th, which does not appear to be too great an excess of strength to allow for wear.

OF THE STRENGTH OF BARS AND RODS TO RESIST TENSION.

293. When the effect of flexure is considered in bars to resist tension, it makes an important difference. Instead of the strength being diminished by flexure, it either has no effect, or has a directly contrary effect. Hence in all works executed in metals the tensile force of the materials should be employed in preference to any other, except the bulk be considerable in respect to the length. In wood we cannot employ a tensile force to much advantage, because it is difficult to form connexions at the extremities of sufficient firmness, but in metals this creates no difficulty.

If a bar or rod be short, its resistance may be computed by the rules, art. 281 and 282.

But when it is long, and the bar is either curved, or the force is not in the direction of the axis, then the effect of flexure may be considered.

The Equation ix. art. 280, will be applicable to all cases where the direction of the force is parallel to the extremities of the bar, that is,

$$\frac{f b d^2}{d + 6 a} = W. \qquad \text{(xviii.)}$$

The flexure is found to be $= \frac{l^2 \epsilon}{4\,d} = \frac{03\,l^2}{d}$, when l is the length in feet, and $\epsilon = \frac{1}{1204}$. But this flexure is to be deducted from the distance from the axis. Hence

$$\frac{15300\,b\,d^3}{d^2 + 6\,a\,d - \cdot 18\,l^2} = \text{W.} \quad \text{[10]} \qquad \text{(xix.)}$$

When the direction of the force is at the distance of half the least side from the axis, then $a = \frac{1}{2}\,d$, and

$$\frac{15300\,b\,d^3}{4\,d^2 - \cdot 18\,l^2} = \text{W.} \qquad \text{(xx.)}$$

And when the direction of the force coincides with the axis,

$$15300\,b\,d = \text{W.} \qquad \text{(xxi.)}$$

When the bar is a cylinder, its strength is to that of a square bar as 1 : 1·6 nearly, (art. 285 ;) hence,

$$\frac{9562\,d^4}{d^2 + 6\,a\,d - \cdot 18\,l^2} = \text{W.} \qquad \text{(xxii.)}$$

Or, when the force is in one of the surfaces of the rod,

$$\frac{9562\,d^4}{4\,d^2 - \cdot 18\,l^2} = \text{W.} \qquad \text{(xxiii.)}$$

It was desirable to show what constituted the advantage of a tensile strain, but I do not intend to

---

[10] In malleable iron,

$$\frac{17800\,b\,d^3}{d^2 + 6\,a\,d - \cdot 16\,l^2} = \text{W.}$$

In oak,

$$\frac{3960\,b\,d^3}{d^2 + 6\,a\,d - \cdot 5\,l^2} = \text{W.}$$

adopt these equations in practical rules, because they are not so simple and easily applied as the rules already given in art. 281 and 282, which will only err a small quantity in excess, when proper care has been taken to take the greatest possible deviation of the straining force from the axis of the piece.

Example 1. Required the weight that may be suspended by a bar of cast iron of 4 inches by 8 inches; under the supposition that the direction of the strain will be in one of the wide surfaces of the bar? Equation xviii. of this art. applies to this case, wherein a is equal 2 inches, or half the least dimension of the bar, that being the distance the direction of the force is supposed to be from the axis; and therefore

$$\frac{f b d^2}{d + 6 a} = \frac{15300 \times 8 \times 4^2}{4 + 12} = 122400 \text{ lbs.}$$

the weight required. See the rule in words at art. 282.

When it is considered that a very small degree of inaccuracy in fitting the connexion may throw the strain all on one side of the bar, the prudence of following this mode of calculation will be apparent.

Example 2. It is proposed to determine the area that should be given to the bars of a suspension bridge, if made of cast iron, for a span of 370 feet; the points of suspension being 30 feet above the lowest point of the curve; and the greatest load, including the weight of the bridge itself, 500 tons.

The load being nearly uniformly distributed, the curve assumed by the chains will not sensibly differ from a parabola;[11] and half the weight will be to the tension at the lower point of the curve, as the rise is to $\frac{1}{4}$th of the span; that is,

$$30 : \frac{370}{4} :: \frac{500}{2} : \frac{500 \times 370}{8 \times 30} = 771 \text{ tons.}$$

This is equal to 1,727,040 ℔s., and by the rule, art. 281, we have

$$\frac{1727040}{15000} = 115 \text{ square inches}$$

for the area of the bars, supposing the stress to be directly in the axis of each; and if we double this area it will provide for a deviation equal to $\frac{1}{6}$th of the diameter of each bar. This will be a sufficient excess of force, considering the great chance of its ever being covered with people, which is the load I have estimated. Hence the sum of the areas of the chains at the lowest point should be 230 square inches. The area at any other point of the curve should be to the area at the lowest point, as the secant of the angle a tangent to the curve makes with a horizontal line, is to the radius. In the present case, the sum of the areas at the point of suspension should be 242 square inches. Cast iron would be greatly superior to wrought iron for chain bridges; it would be more durable, less expensive to obtain the same strength, and when made suf-

[11] See Elementary Principles of Carpentry, art. 57.

ficiently strong, its weight would prevent excessive vibration by small forces. Most of the wrought iron bridges appear to be very slight and temporary structures when examined by the rules I have given, which appear to be founded on unquestionable principles.

Example 3. To determine the area of a piston-rod for a single-acting engine, the force on the piston being equivalent to 11 lbs. on a square inch, and allowing for the possibility of the direction of the force being at half the diameter of the rod from its axis. In this case 11 times the square of the diameter of the piston in inches is equal to the stress, and if D be the diameter of the steam piston, and d that of the piston-rod, we have for wrought iron,

$$3{\cdot}1416 \times 11\,D^2 = \frac{3{\cdot}1416\,d^2 \times 17800}{4};$$

or very nearly,

$$\frac{D}{20} = d.$$

That is, the diameter of the piston-rod should be $\frac{1}{20}$th of the diameter of the steam cylinder, when nothing is allowed for wear; or making the allowance which appears to be requisite, the diameter should be $\frac{1}{15}$th of the diameter of the steam cylinder.

SECTION XI.

OF THE STRENGTH OF CAST IRON TO RESIST AN IMPULSIVE FORCE.[1]

294. The moving force of a body, or of a part of a machine, ought to be balanced by the elastic force of the parts which propagate the motion; for if the effect of the moving force be greater than the elastic force of the parts, some of them will ultimately break; besides, a part of the power of the machine will be lost at each stroke.

And since increasing the mass of matter to be moved increases the friction in a machine, it is an advantage to employ no more material in its moving parts than is absolutely necessary for strength; but, in other parts exposed to pressive forces only, it is desirable that the materials should always be capable of resisting the strains, with as small a degree of flexure as is convenient, because steadiness is, in

---

[1] Dr. Young has given the term *resilience* to this species of resistance; and the reader will find some interesting remarks on the importance of studying it, in his Lectures on Nat. Phil. vol. i. p. 143.

the fixed parts of machines, a most desirable property.

A beam resists a moving force, as a spring, by yielding and opposing the force as it yields, till it finally overbalances it;[2] and hence it is, that a brittle or very stiff body breaks, because it does not yield sufficiently for destroying the force.

As the resistance of a beam under different degrees of flexure can be calculated, the effect of that resistance in the destruction of motion may be estimated by the principles of dynamics: such inquiries are usually managed by the method of fluxions; but not being satisfied with the manner of establishing the principles of that method, though I have no doubt of the correctness of results obtained by it, I shall briefly deduce the rules of this section by another mode of calculation.

295. If the intensity of a force be variable, so that the action upon the body moved at any point be directly as some power, n, of the distance from a point B, fig. 23, towards which it moves. Then, if the intensity of the force at A be equal P, the intensity at any point C will be $\frac{(CB)^n . P}{(AB)^n}$. For, by the definition,

$$(AB)^n : (CB)^n :: P : \frac{(CB)^n . P}{(AB)^n}.$$

Put S to denote the space AB; and conceive this

---

[2] For a machine to produce the greatest effect, the time of bending the beam should be as small as possible.

space S to be divided into m equal parts, denoting any one of these parts by x; and in consequence of the smallness of these parts, if we take the mean between the intensity at the beginning, and that at the end of each part, and consider each of these means an uniform intensity for the space it was calculated for, then these uniform intensities may be represented by the following progression:

$$\frac{P}{2S^n}\times\left(\overline{0+x^n}+\overline{x^n+2^n x^n}+\overline{2^n x^n+3^n x^n}+\ldots\overline{\overline{m-1}^n x^n+m^n x^n}\right)$$

or,

$$\frac{P x^n}{S^n}\left\{1^n+2^n+3^n+\ldots\overline{m-1}^n+\frac{m^n}{2}\right\}.$$

296. It is shown by writers on dynamics, that when the intensity of a force is uniform, the square of the quantity of force accumulated or destroyed is directly as the intensity multiplied by the quantity of matter moved, and by the space moved through.[3] Therefore, making W = the quantity of matter, and g a constant quantity to reduce the proportion to an equation, we find the square of the forces accumulated or destroyed, in the space S, may be exhibited by the progression

$$\frac{g P W x^{n+1}}{S^n}\left\{1^n+2^n+\ldots\overline{m-1}^n+\frac{m^n}{2}\right\}.$$

And, from the principles of the method of progressions,[4] the accurate value of the square of the force accumulated or destroyed in the space S is

---

[3] Dr. C. Hutton's Course of Math. vol. ii. p. 136. 5th edit.

[4] See Philosophical Magazine, vol. lvii. p. 201.

$$\frac{g\,\mathrm{P\,W\,S}}{n+1}.$$

297. When $n = o$, or the intensity is uniform, the square of the accumulated force is $= g\,\mathrm{PWS}$.

298. The force of gravity near the earth's surface is nearly uniform, and in this case we know from experiments on falling bodies that $g = 64\frac{1}{3}$, and $\mathrm{P}=\mathrm{W}$ the weight of the body; therefore, $64\frac{1}{3}\,\mathrm{W}^2\,\mathrm{S}$ = the square of the accumulated force, and $64\frac{1}{3}$ may be substituted for g.

Hence the moving force of a falling body is $\mathrm{W}\sqrt{64\frac{1}{3}\,\mathrm{S}}$.

299. If $n = 1$, we have

$$\frac{g\,\mathrm{P\,W\,S}}{n+1} = \frac{64\frac{1}{3}\,\mathrm{P\,W\,S}}{2} = 32\tfrac{1}{6}\,\mathrm{P\,W\,S};$$

and as, in the resistance of beams, the intensity at any deflexion is directly as the deflexion, the quantity $32\frac{1}{6}$ PWS represents the square of the force destroyed in producing a deflexion equal to S. That is, when a beam is supported at both ends, and S = the deflexion in the middle, in decimal parts of a foot, then $\sqrt{32\frac{1}{6}\,\mathrm{PWS}}$ = the force that would be destroyed in producing the flexure S; where P is the weight that would produce the deflexion S. [5]

---

[5] The effect of elastic gases in producing or destroying motion is expressed by the same equation, when the change of bulk is not so rapid as to cause cold in the one case, or to develope heat in the other. The developement of heat by the sudden compression of air materially affects the velocity of sound, and was first applied by Laplace to correct the discrepancy between theory

Having considered the effect of the resisting force of the material in destroying an impulsive force, we must now consider the circumstances which take place in the different cases occurring in practice.

300. If the blow be made by a falling body in the direction of gravity, and the weight of the falling body be w, and its velocity at the time of impact be v, then by the laws of collision, in the case of equilibrium,

$$v w = \sqrt{32\tfrac{1}{6}\,\mathrm{P\,S}\,(\mathrm{W} + w)}. \qquad \text{(i.)}$$

In which equation the small acceleration that would be produced by the action of gravity on the mass $\mathrm{W} + w$, during the flexure of the beam, is neglected.

301. If the blow were made horizontally by a body of the weight w, moving with a velocity v, then the equation is correct; and even in the first case it is accurate enough for practical purposes.

302. If the blow were made by a weight w falling from a given height h, we have by the laws of gravity, (art. 298,)

$$w\,v = w\,\sqrt{64\tfrac{1}{3}\,h};$$

therefore,

$$w\,\sqrt{64\tfrac{1}{3}\,h} = \sqrt{32\tfrac{1}{6}\,\mathrm{P\,S}\,(\mathrm{W} + w)}, \text{ or}$$

$$2\,w^2\,h = \mathrm{P\,S}\,(\mathrm{W} + w). \qquad \text{(ii.)}$$

303. When the strain is occasioned by a force of

---

and experiments; a subject which has been further illustrated by the researches of Poisson, in an article "Sur la Vitesse du Son," Annales de Chimie, tome xxiii. p. 5.

an intensity F, and velocity v, such for example as would be occasioned by the sudden derangement of a machine in motion with the velocity v, and force F, then

$$F v = \sqrt{32\tfrac{1}{6} \, P S W}; \text{ or,}$$

$$F^2 v^2 = 32\tfrac{1}{6} \, P S W. \qquad \text{(iii.)}$$

The last equation is applicable to the beams of steam engines, and in general to reciprocating movements in machines, such as the connecting rods, cranks, &c.

If a body be previously in motion in the direction of the impulsive force, then the force $F v$ should be the difference between the forces of the impelling and impelled bodies.

304. A general number of comparison to exhibit the power of a body to resist impulse, and which might be termed the *modulus of resilience*, would be extremely convenient in calculations of this kind; and when we omit the effect of a difference of density, which it is usual to do, we have an easy method of forming such a number.[6] For in any case, if f

---

[6] The number might include the effect of density, if we were to measure the resistance to impulse by the height a body should fall to produce permanent change by its own weight; for we easily derive from Equation ii. art. 302,

$$h : \frac{f \epsilon}{s},$$

when s is the specific gravity. This expression might be termed the *specific resilience* of a body, and if it were denoted by Σ, we should have

be the force which produces permanent alteration, and ϵ the corresponding extension,

$$PS : f\epsilon.$$

And, since in bars of different materials placed in the same circumstances the resistance to impulse may be considered proportional to the height a body must fall to produce a permanent change in the structure of the matter; and as that height is proportional to PS, and consequently to $f\epsilon$, when the effect of density is neglected; we may take $f\epsilon$ the measure of the power of a body to resist impulsion, that is, the modulus of resilience; and representing this modulus by R,

$$f\epsilon = R. \qquad \text{(iv.)}$$

In cast iron,

$$f = 15300; \text{ and } \epsilon = \frac{1}{1204};$$

$$\text{therefore } R = 12{\cdot}7.$$

305. These equations flow from the principle that while the elasticity is perfect, the deflexion or extension is as the force producing it, but it also varies according to the manner in which the material is strained. In some cases, of frequent occurrence, the application is shown in the examples.

But it will be useful, before we proceed any further, to inquire what velocity cast iron will bear,

---

$$\frac{f\epsilon}{s} = \Sigma.$$

In cast iron,

$$\Sigma = 1{\cdot}762.$$

without permanent alteration, in order that we may be aware whether such velocity will ever take place in the parts of machines; for if any part of a machine be connected with others that will yield to the force, and the material be capable of transmitting the motion with greater velocity than the machine moves with, it need be formed only for resistance to power or pressure.

306. It has been shown that $\sqrt{32\frac{1}{6}\,\mathrm{P\,W\,S}}$ is equal to the greatest force an elastic body can generate or destroy (art. 299); if it were exposed to a greater force, its arrangement would be permanently altered. Now, if V be the greatest velocity the body is capable of transmitting, if communicated to its mass, we have

$$\sqrt{32\tfrac{1}{6}\,\mathrm{P\,W\,S}} = \mathrm{V\,W}, \text{ or}$$

$$\sqrt{\frac{32\frac{1}{6}\,\mathrm{P\,S}}{\mathrm{W}}} = \mathrm{V}. \qquad \text{(v.)}$$

307. It has also been shown, that in cast iron the cohesive force $f = 15{,}300$ ℔s. (art. 143), and the extension

$$\epsilon = \frac{1}{1204} \text{ (art. 212)};$$

and since $\mathrm{S} = l\,\epsilon$, (by art. 104,) and $\mathrm{P} = b\,d\,f$, (art. 103,) and $l\,b\,d\,p = \mathrm{W}$, where $p = 3{\cdot}2$ ℔s., the weight of a bar of iron 12 inches long and 1 inch square; therefore, when a bar is strained in the direction of its length,

$$\sqrt{\frac{32\frac{1}{6}\,\mathrm{P\,S}}{\mathrm{W}}} = \sqrt{\frac{32\frac{1}{6} \times b\,d\,f \times l\,\epsilon}{l\,b\,d\,p}} = \sqrt{\frac{32\frac{1}{6} \times 15300}{3{\cdot}2 \times 1204}} = \mathrm{V} =$$

11·3 feet per second.

308. If an uniform bar be supported at the ends, we have

$$P = \frac{850\, b\, d^2}{l} \text{ (art. 143) and } S = \frac{\cdot 02\, l^2}{12\, d} \text{ ft. (art. 212)};$$

also,

$$W = \frac{l\, b\, d\, p}{2},$$

for the mass of the beam would acquire only the same momentum as half of it collected in the middle. Consequently,

$$\sqrt{\frac{32\frac{1}{6}\, P\, S}{W}} = \sqrt{\frac{32\frac{1}{6} \times 850 \times \cdot 02 \times 2}{12 \times 3\cdot 2}} = V =$$

5·3366 feet per second, nearly.

I have shown by a comparison of many experiments in art. 70, that about 3·3 times the force that produces permanent alteration will break a beam; therefore, assuming the deflexion to continue proportional to the force till fracture takes place, we have

$$\sqrt{\frac{32\frac{1}{6} \times 3\cdot 3\, P \times 3\cdot 3\, S}{W}} = V; \text{ or}$$

$$3\cdot 3 \sqrt{\frac{32\frac{1}{6}\, P\, S}{W}} = V.$$

Therefore, a velocity of

$$3\cdot 3 \times 5\cdot 3366 = 17\cdot 6 \text{ feet per second,}$$

would break a beam; or a beam would break by falling a height of about 5 feet.

309. Hence it is clear, that cast iron is capable of sustaining only a very small degree of velocity; and a correct knowledge of this limit is certainly of the

first importance in the application of this material in machinery. When a cast iron bar is exposed to an impulsive force in the direction of its length, the utmost velocity its mass should acquire must never exceed 11 feet per second; and when the force acts in a direction perpendicular to the length, it should never be capable of communicating to the mass of the bar a greater velocity than about 5 feet per second; and if it exceed 18 feet per second, the bar will break.

If the connecting rod of a steam engine were to move with a greater velocity than 5 feet per second, the swag of its own weight would produce permanent flexure.

If a ship with hollow cast iron masts should strike a rock when it moved with a velocity of 12 miles per hour, the masts would break; and even with less velocity, for here we neglect the effect of the wind on the masts.

310. To illustrate the use of the above investigation, or rather, to prevent any one from disappointment, in applying these rules for the resistance to impulsion, it may be useful to consider how they should be applied to the parts of machines. In a machine, the motion is communicated from the impelled to the working point by a certain number of parts, and among these parts one at least should be capable of resisting the whole energy of the moving power. If there be many parts to transmit the power, then two or more of them should be

capable of resisting the energy of the moving power, and they should be distributed so as to divide the line of communication into nearly equal parts. If the intermediate parts be made sufficient to resist the dead power of the machine, that is, the power without velocity, they will always be strong enough to convey the velocity, if it be less than is stated in the preceding article, to other parts, that will either forward it to the working point, or resist it entirely during a momentary derangement of the action of the machine. To make all the parts strong enough for this purpose would often cause a machine to be clumsy, and unfit for any practical use.

311. Let the constant numbers for the strength and deflexion in feet be $f\,\delta$. Then,

$$P = \frac{f\,b\,d^2}{l}, \text{ and } S = \frac{\delta\,l^2}{d}.$$

Also, let the weight of the beam itself be n times the weight of the falling body. These values being substituted in Equation i. art. 300, we have

$$v\,w = \sqrt{32\tfrac{1}{6}\,P\,S\,\overline{W + w}} = \sqrt{32\tfrac{1}{6}\,l\,b\,d\,f\,\delta\,w\,(n+1)}\text{; or,}$$

$$\frac{v^2\,w}{32\tfrac{1}{6}\,l\,f\,\delta\,(n+1)} = b\,d. \qquad \text{(vi.)}$$

312. If the like substitutions be made in Equation iii. art. 303, we obtain

$$F^2\,V^2 = (32\tfrac{1}{6}\,P\,S\,W) = 32\tfrac{1}{6}\,l\,b\,d\,f\,\delta\,W\text{;}$$

and if $l\,b\,d\,p$ be the weight of the mass of the beam the force acts upon, then

$$\frac{F\,V}{l\sqrt{32\tfrac{1}{6}\,f\,\delta\,p}} = b\,d. \qquad \text{(vii.)}$$

PRACTICAL RULES AND EXAMPLES.

313. *Prop.* I. To determine a rule for finding the dimensions of a beam to resist the force of a body in motion.

It is evident by Equation vi. art. 311, that the error which would arise from neglecting to allow for the effect of the weight of the beam itself, would always be on the safe side in calculating the dimensions of a beam to resist an impulsive force; and since, by such neglect, the rule is reduced to a very simple form, instead of a very complicated one, I shall apply the equation under the form

$$\frac{v^2 w}{32\frac{1}{6}\, l f \delta} = b\, d.$$

314. *Case* 1. When the beam is uniform and supported at the ends. In this case $f = 850$, (see art. 143,) and δ in feet $= \frac{\cdot 02}{12}$ (by art. 212,) hence,

$$32\tfrac{1}{6} f \delta = 45{\cdot}5 \text{; or}$$

$$\frac{v^2 w}{45{\cdot}5\, l} = b\, d.$$

315. *Rule.* Multiply the weight of the falling body in pounds by the square of its velocity in feet per second; divide this product by 45·5 times the length in feet, and the quotient will be the area in inches.

The depth should be at least sufficient to render the beam capable of supporting its own weight, added to the weight of the falling body, which may be readily found by Table II. art. 6.

316. If the height of the fall be given instead of the velocity of the falling body, then instead of multiplying by the square of the velocity, multiply by sixty-four times the height of the fall.

317. *Example* 1. To determine the area of a cast iron beam that would sustain, without injury, the shock of a weight of 170 lbs. falling upon its middle with a velocity of 8 feet per second, the distance between the supports being 26 feet. By the rule

$$\frac{170 \times 8^2}{45 \cdot 5 \times 26} = 9 \cdot 2 \text{ inches, the area required.}$$

Hence, if we make the depth 6 inches, the breadth will be 1·53 inches, and the beam would sustain a pressure of 1800 lbs. (see Table II.) to produce the same effect as the fall of 170 lbs. It may also be observed, that half the weight of the beam is 400 lbs., making 570 lbs. for the pressure the beam would have to sustain after the velocity was destroyed, which is not quite $\frac{1}{3}$rd of the weight the beam would bear.

318. *Example* 2. If a bridge of 30 feet span were formed on beams of cast iron, of what area should the section of these beams be, so that any one of them might be sufficient to resist the impulsive force of a waggon wheel falling over a stone 3 inches high, the load upon that wheel being 3360 lbs.?

The height of the fall being ·25 foot, the square of the velocity acquired by the fall will be 64 × ·25 = 16; therefore,

$$\frac{3360 \times 16}{45 \cdot 5 \times 30} = 39 \cdot 338 \text{ inches,}$$

the area required.

This area is nearly 40 inches; suppose it 40, then 40 × 15 × 3·2 = 1920 lbs. = half the weight of the beam, (that is, the area, in inches, multiplied by half the length in feet, multiplied by 3·2 lbs., the weight of a piece of cast iron, 1 foot in length, and 1 inch square;) consequently, 1920 + 3360 = 5280 lbs., the whole effective pressure on the beam, after the velocity is destroyed. If we were to make the beam 20 inches deep, and 2 inches in thickness, it may be found by Table II. that the deflexion would be ·9 of an inch, and it would require a pressure of 45,328 lbs. to produce the same effect as the fall of the wheel, above eight times the pressure of the load and weight.

319. *Case* 2. When a beam is supported at the ends, the breadth uniform, and the outline of the depths an ellipse.

This case applies to bridges or beams to withstand an impulsive force at any point of the length. By art. 144, $f = 850$, and by art. 139,

$$\delta \text{ in feet} = \frac{\cdot 0257}{12};$$

therefore the equation

$$\frac{v^2 w}{32\frac{1}{6} f \delta l} = b\, d,$$

in art. 313, becomes

$$\frac{v^2 w}{58{\cdot}5\, l} = b\, d.$$

320. *Rule.* Calculate by the rule, art. 315, with 58·5 as a divisor instead of 45·5.

321. *Case* 3. When the breadth and depth of a beam are uniform, and the section is as fig. 9, Plate I., and the beam supported at the ends.

In this case $f = 850\,(1 - q\,p^3)$ by art. 186, and

$$\delta = \frac{{\cdot}02}{12}\text{ foot}$$

by art. 212; hence the equation (art. 313),

$$\frac{v^2 w}{32\frac{1}{6} f \delta\, l} = \frac{v^2 w}{45{\cdot}5\,(1 - p^3\, q)\, l} = b\, d;$$

consequently, the power of a beam to resist an impulsive force, when the quantity of material is the same, is considerably increased by giving this form to the section.

322. *Case* 4. If a beam of the form of section shown in fig. 9, be the elliptical form of equal strength, (see fig. 24, Plate III.) then

$$\frac{v^2 w}{58{\cdot}5\, l\,(1 - p^3\, q)} = b\, d,$$

when the beam is supported at both ends, and the impulsive force acts at any point of the length.

323. *Case* 5. In an open beam, as fig. 11, Plate II., we may consider the beam as bounded by a semi-ellipse, when the breadth is uniform, and in this case

$$\frac{v^2 w}{58{\cdot}5\, l\,(1 - p^3)} = b\, d.$$

324. *Example.* To determine the area of the section of an open girder, that would sustain the shock of 300 lbs. falling from a height of 1 foot, the length between the supports being 26 feet, and the depth of the open part $\frac{7}{10}$ of the whole depth.

In this example

$$\frac{v^2 w}{58{\cdot}5\, l\,(1 - p^3)} = \frac{64 \times 300}{58{\cdot}5 \times 26\,(1 - {\cdot}343)} =$$

20 inches nearly. This is perhaps as great an impulsive force as it is probable a girder for a room will be likely to be exposed to; and since this area of section would not be sufficient for the greatest pressure, it appears unnecessary to calculate the effect of moving force in the construction of girders.

325. *Prop.* II. To determine a rule for finding the dimensions of an uniform beam to resist a moving force.

This proposition applies to the parts of machines, and as there are few people engaged in the construction of powerful machines that are not competent to apply an equation, I shall in this part give the rules in the form of equations only.

326. *Case* 1. When an uniform beam is supported at the ends, and the moving force acts at the middle of the length.

By art. 143, $f = 850$, and by art. 212,

$$\delta = \frac{{\cdot}02}{12} = {\cdot}0016\dot{6} \text{ foot};$$

and since 3·2 lbs. = the weight of 1 foot in length, and 1 inch square, we shall have

$$p = \frac{3·2}{2} = 1·6;$$

therefore,

$$\frac{FV}{l\sqrt{33\frac{1}{6} f \delta p}} \text{ (art. 312)} = \frac{FV}{l\sqrt{32\frac{1}{6} \times 850 \times ·001\dot{6} \times 1·6}} = \frac{FV}{8·6\, l} = b\, d.$$

327. *Rule.* When F is the force in pounds, V its velocity in feet per second, l the whole length in feet between the supports, b the breadth, and d the depth in inches, then

$$\frac{FV}{8·6\, l} = b\, d.$$

328. *Case* 2. When an uniform beam rests upon a centre of motion, and the moving force acts at one end, and is opposed by a greater resistance at the other end.

By art. 153,

$f = 212$, and by art. 220, $\delta = ·08\ (1 + r,)$ and $p = \frac{3·2}{2}$; hence,

$$\frac{FV}{l\sqrt{32\frac{1}{6} f \delta p}} = \frac{FV}{8·6\, l\sqrt{1 + r}} = b\, d.$$

329. *Rule.* Make F = the force in pounds, V its velocity in feet per second, l = the length in feet between the centre of motion and the point where the force acts, and l' = the length in feet between the centre of motion and the point of resistance; b and d being the breadth and depth in inches; then

$$\frac{l'}{l} = r, \text{ and } \frac{\text{F V}}{8{\cdot}6\, l \sqrt{1 + r}} = b\, d.$$

330. If $l = l'$ we have

$$\frac{\text{F V}}{8{\cdot}6\, l \sqrt{2}} = \frac{\text{F V}}{12{\cdot}2\, l} = b\, d.$$

331. *Example.* To determine the area of the section of the beam for a steam engine, when it is to be of uniform depth; the length 24 feet, the centre of motion in the middle of the length; the pressure upon the piston 5000 ℔s., and its greatest velocity 4 feet per second.

By art. 330,

$$\frac{\text{F V}}{12{\cdot}2\, l} = b\, d = \frac{5000 \times 4}{12{\cdot}2 \times 12} = 137 \text{ inches nearly.}$$

If this beam were made 30 inches deep, the deflexion by such a strain would be about $\frac{8}{10}$ths of an inch, and the breadth would be

$$\frac{137}{30} = 4{\cdot}57 \text{ inches,}$$

and such a beam would bear a weight of about 12 times the pressure on the piston, without destroying its elastic force.

332. *Prop.* III. To determine a rule for finding the area of the middle section of a parabolic beam to resist a moving force when the breadth is uniform.

The motion communicated to the arm of a lever is the same as if its whole weight were collected at its centre of gravity; and as the length of the arm is to the distance of its centre of gravity, so is the mass to the effect of that mass collected at the extremity. Therefore, when the distance of the centre of gravity is some part of the length, the effect of the mass of the arm will be the same part of the whole of its weight when acting at the extremity.

333. *Case* 1. When a parabolic beam is supported at both ends, and the moving force acts at the middle of the length.

By art. 143, $f = 850$, and by art. 224,

$$\delta = \frac{\cdot 04}{12} = \cdot 003\dot{3} \text{ foot.}$$

Also, because the area of a parabolic beam is $\frac{2}{3}$ of one uniformly deep,[7] and the distance of the centre of gravity from the centre of motion is $\frac{3}{5}$ of the length;[8] we have

$$p = 3\cdot 2 \times \frac{2}{3} \times \frac{3}{5} = \frac{6\cdot 4}{5} = 1\cdot 28.$$

Consequently, the equation, (art. 312,)

$$\frac{FV}{l\sqrt{32\frac{1}{6} f \delta p}} = \frac{FV}{l\sqrt{32\frac{1}{6} \times 850 \times \cdot 003\dot{3} \times 1\cdot 28}} = \frac{FV}{l \times 10\cdot 8} = bd.$$

334. In beams supported at both ends, and of the same breadth, the power of a parabolic beam to resist a moving force, is to that of an uniform beam, as 10 is to 8 nearly; and the parabolic beam requires very little more than $\frac{2}{3}$rds of the quantity of material.

---

[7] Dr. Hutton's Course, vol. ii. p. 126.

[8] Idem, vol. ii. p. 327.

335. *Rule.* When F is the force in pounds, V its velocity in feet per second, l the whole length between the supports, and b and d the breadth and depth in inches; then

$$\frac{FV}{10\cdot 8\,l} = b\,d.$$

336. *Example.* Let the force of a steam engine be applied to the middle of its beam, so as to cause it to move an axis by means of two cranks, placed so as to be impelled by the ends of the beam. Let the greatest pressure on the piston be 3000 ℔s., its greatest velocity 3 feet per second, and the whole length 12 feet.

By the rule, (art. 335,)

$$\frac{FV}{10\cdot 8\,l} = \frac{3000 \times 3}{10\cdot 8 \times 12} = b\,d = 70 \text{ inches.}$$

337. *Case* 2. When a parabolic beam rests upon a centre of motion, and the moving force acts at one end, and is opposed by a greater resistance at the other.

By art. 153, $f = 212$, and by art. 227,

$$\delta = \frac{16\,(1+r)}{12};$$

also,

$$p = 32 \times \frac{2}{3} \times \frac{2}{5} = \frac{12\cdot 8}{15} = \cdot 8533\dot{3};$$

hence,

$$\frac{FV}{l\sqrt{32\frac{1}{6}\,f\,\delta\,p}} = \frac{FV}{l\sqrt{32\frac{1}{6} \times 212 \times \cdot 85\dot{3} \times \frac{\cdot 16\,(1+r)}{12}}} = \frac{FV}{8\cdot 82\,l\sqrt{1+r}} = bd.$$

338. *Rule.* Make F the force in pounds, and V its velocity in feet per second, $l =$ the length in

feet, from the centre of motion to the point where the force acts, and l' the length from the centre of motion to the resisted point; also, make b and d the breadth and depth in inches; then

$$\frac{l'}{l} = r, \text{ and } \frac{\text{F V}}{8{\cdot}82\, l \sqrt{1 + r}} = b\, d.$$

339. If $l = l'$, that is, when the centre of motion is in the middle of the beam,

$$\frac{\text{F V}}{12{\cdot}5\, l} = b\, d.$$

340. In a steam engine the weight of the connecting apparatus, the power applied to the air-pump, and the weight of the catch-pins, should be allowed for; and when the engine moves machinery, the beam should not be less than is determined by this rule. The depth of the beam is usually the same as the diameter of the steam piston.

341. *Example.* If the pressure on the piston of a steam engine be 15,000 lbs., the whole length of the beam 24 feet, and its velocity 3 feet per second, required the area of the beam?

In this case,

$$\frac{\text{F V}}{12{\cdot}5\, l} = \frac{15000 \times 3}{12{\cdot}5 \times 12} = b\, d = 300 \text{ inches.}$$

If the beam be made 48 inches deep, it should be $6\frac{1}{4}$ inches in breadth; and the best method of forming such a beam is to make it in two parts, each $3\frac{1}{8}$ in breadth, placed at 12 or 14 inches apart, and well connected together. This arrangement causes an engine to work with more steadiness, and the parts are less troublesome to move and fix in their places than a single mass would be.

342. *Prop.* IV. To determine a rule for finding the area of the middle section of a beam of uniform breadth, the depth at the end being half the depth in the middle, and the middle of the depth open, to resist a moving force.

Let the parts be so arranged that the centre of gravity may be considered to be at the middle of the length of the arm of the beam, which will be very nearly true in practice, and will render the computation somewhat easier.

343. *Case* 1. When an open beam is supported at the ends, and the force is applied in the middle of the length.

By art. 200, $f = 850\ (1 - p'^3)$, and by art. 231,

$$\delta = \frac{\cdot 0327}{12} = \cdot 002725\,;$$

also, we have

$$p = \frac{3\cdot 2 \times (1 - p')}{2} = 1\cdot 6\ (1 - p')\,;$$

and the Equation, (art. 312,)

$$\frac{\mathrm{F\,V}}{l\sqrt{32\frac{1}{6} f \delta p}} = \frac{\mathrm{F\,V}}{l\sqrt{32\frac{1}{6} \times 850\,(1 - p'^3) \times \cdot 002725 \times 1\cdot 6\,(1 - p')}} =$$

$$\frac{\mathrm{F\,V}}{10\cdot 92\, l \sqrt{(1 - p'^3) \times (1 - p')}} = b\,d.$$

344. *Rule.* Make F the force in pounds, V its velocity in feet per second, l the whole length be-

tween the supports in feet, p' that number which would be produced by dividing the depth of the part left out in the middle, by the whole depth; (if this ratio were not fixed, the solution could not be effected;) and b and d the breadth and depth in the middle in inches; then

$$\frac{\mathrm{F\,V}}{10{\cdot}92\,l\,\sqrt{(1-p'^3)\times(1-p')}} = b\,d.$$

345. If p' be made $= {\cdot}7$, which is a convenient proportion, then

$$\frac{\mathrm{F\,V}}{4{\cdot}85\,l} = b\,d.$$

346. *Case* 2. When an open beam is supported on a centre of motion, and the moving force acts at one end, and the resistance at the other.

By the same method as above we find

$$\frac{\mathrm{F\,V}}{10{\cdot}92\,l\,\sqrt{(1-p^3)\times(1-p)\times(1+r)}} = b\,d.$$

347. *Rule.* Make F = the force in pounds, V its velocity in feet per second, l the length from the point where the force acts to the centre of motion in feet, and l' the length from the centre of motion to the point of resistance, b and d the breadth and depth in inches in the middle of the beam, and p the number arising from the division of the depth of the part left out in the middle by the whole depth; then, $\frac{l'}{l} = r$, and

$$\frac{FV}{10{\cdot}92\, l \sqrt{(1-p^3) \times (1-p) \times (1+r)}} = b\,d.$$

348. If $p = {\cdot}7$, the equation reduces to

$$\frac{FV}{4{\cdot}85\, l \sqrt{1+r}} = b\,d.$$

349. Also, when the centre of motion is in the middle of the beam, and $p = {\cdot}7$, we have

$$\frac{FV}{6{\cdot}86\, l} = b\,d.$$

350. *Example.* As an example to the equation in the last article, let us suppose the pressure on the piston of a steam engine to be 15,000 ℔s., its velocity 3 feet per second, and the whole length of the beam 24 feet, which is the same as the example (art. 341). In this case

$$\frac{FV}{6{\cdot}86\, l} = \frac{15000 \times 3}{6{\cdot}86 \times 12} = 771 \text{ inches} = b\,d.$$

And, if the depth be made 48 inches, then

$$\frac{771}{48} = 16{\cdot}06$$

inches the breadth, which is to be the same throughout the length. The bulk of the metal in the upper and lower part of the beam will be found by multiplying the depth by ·7; that is, $\cdot 7 \times 48 = 33{\cdot}6$; which, deducted from 48, leaves 14·4 inches, or 7·2 inches for each side.

Fig. 34, Plate IV., shows a sketch for a beam of this kind, drawn according to these proportions.

351. Any of the rules of this, or of the preceding Section, may be applied to other materials by substituting the proper values of the cohesive force, extensibility, and density; these are given for several kinds in the following Table.

TABLE OF DATA,[1] &c.

USEFUL IN VARIOUS CALCULATIONS;

ARRANGED ALPHABETICALLY.

The data correspond to the mean temperature and pressure of the atmosphere, dry materials; and the temperature is measured by Fahrenheit's scale.

AIR. Specific gravity 0·0012; weight of a cubic foot 0·0753 ℔., or 527 grains, (SHUCKBURGH); 13·3 cubic feet or 17 cylindric feet of air weigh 1 ℔.; it expands $\frac{1}{480}$ or ·00208 of its bulk at 32° by the addition of one degree of heat (DULONG and PETIT.)

ASH. Specific gravity 0·76; weight of a cubic foot 47·5 ℔s.; weight of a bar 1 foot long and 1 inch

---

[1] I have left this part, like the body of the work, unaltered, though the results in it sometimes differ much from my own; and beg to refer the reader for information derived from experiment to the 'Additions,' and particularly to a Paper of mine on the ultimate Tensile, Compressive, and Transverse Strength of various kinds of materials, including Timber of the sorts in most general use, Building Stone, Marble, Slate, Glass, &c.; which will be offered to the Royal Society in a very short time.—EDITOR.

square 0·33 lb.; will bear without permanent alteration a strain of 3540 lbs. upon a square inch, and an extension of $\frac{1}{464}$ of its length; weight of modulus of elasticity for a base of an inch square 1,640,000 lbs.; height of modulus of elasticity 4,970,000 feet; modulus of resilience 7·6; specific resilience 10. (Calculated from BARLOW's Experiments.)

Compared with cast iron as unity, its strength is 0·23; its extensibility 2·6; and its stiffness 0·089.

ATMOSPHERE. Mean pressure of, at London, 28·89 inches of mercury = 14·18 lbs. upon a square inch. (ROYAL SOCIETY.) The pressure of the atmosphere is usually estimated at 30 inches of mercury, which is very nearly 14¾ lbs. upon a square inch, and equivalent to a column of water 34 feet high.

BEECH. Specific gravity 0·696; weight of a cubic foot 45·3 lbs.; weight of a bar 1 foot long and 1 inch square 0·315 lb.; will bear without permanent alteration on a square inch 2360 lbs., and an extension of $\frac{1}{570}$ of its length; weight of modulus of elasticity for a base of an inch square 1,345,000 lbs.; height of modulus of elasticity 4,600,000 feet; modulus of resilience 4·14; specific resilience 6. (Calculated from BARLOW's Experiments.)

Compared with cast iron as unity, its strength is 0·15; its extensibility 2·1; and its stiffness 0·073.

BRASS, *cast*. Specific gravity 8·37; weight of a cubic foot 523 lbs.; weight of a bar 1 foot long and 1 inch square 3·63 lbs.; expands $\frac{1}{93800}$ of its length by one degree of heat (TROUGHTON); melts at 1869° (DANIELL); cohesive force of a square inch 18,000 lbs. (RENNIE); will bear on a square inch without permanent alteration 6700 lbs., and an extension in length of $\frac{1}{1333}$; weight of modulus of elasticity for a base of an inch square 8,930,000 lbs.; height of modulus of elasticity 2,460,000 feet; modulus of resilience 5; specific resilience 0·6 (TREDGOLD).

Compared with cast iron as unity, its strength is 0·435; its extensibility 0·9; and its stiffness 0·49.

BRICK. Specific gravity 1·841; weight of a cubic foot 115 lbs.; absorbs $\frac{1}{15}$ of its weight of water; cohesive force of a square inch 275 lbs. (TREDGOLD); is crushed by a force of 562 lbs. on a square inch (RENNIE.)

BRICK-WORK. Weight of a cubic foot of newly built, 117 lbs.; weight of a rod of new brick-work 16 tons.

BRIDGES. When a bridge is covered with people, it is about equivalent to a load of 120 lbs. on a superficial foot; and this may be esteemed the greatest possible extraneous load that can be collected on a bridge; while one incapable of supporting this load cannot be deemed safe.

BRONZE. See Gun-metal.

CAST IRON. Specific gravity 7·207; weight of a cubic foot 450 lbs.; a bar 1 foot long and 1 inch square weighs 3·2 lbs. nearly; it expands $\frac{1}{162000}$ of its length by one degree of heat (ROY); greatest change of length in the shade in this climate $\frac{1}{1723}$; greatest change of length exposed to the sun's rays $\frac{1}{1270}$; melts at 3479° (DANIELL), and shrinks in cooling from $\frac{1}{98}$ to $\frac{1}{85}$ of its length (MUSCHET); is crushed by a force of 93,000 lbs. upon a square inch (RENNIE); will bear without permanent alteration

15,300 ℔s.[2] upon a square inch, and an extension of $\frac{1}{1204}$ of its length; weight of modulus of elasticity for a base of an inch square 18,400,000 ℔s.; height of modulus of elasticity 5,750,000 feet; modulus of resilience 12·7; specific resilience 1·76 (TREDGOLD).

CHALK. Specific gravity 2·315; weight of a cubic foot 144·7 ℔s.; is crushed by a force of 500 ℔s. on a square inch. (RENNIE.)

CLAY. Specific gravity 2·0; weight of a cubic foot 125 ℔s.

COAL. *Newcastle.* Specific gravity 1·269; weight of a cubic foot 79·31 ℔s. A London chaldron of 36 bushels weighs about 28 cwt., whence a bushel is 87 ℔s. (but is usually rated at 84 ℔s.) A Newcastle chaldron, 53 cwt. (SMEATON.)

COPPER. Specific gravity 8·75 (HATCHETT); weight of a cubic foot 549 ℔s.; weight of a bar 1 foot long and 1 inch square 3·81 ℔s.; expands in length by one degree of heat $\frac{1}{105900}$ (SMEATON); melts at 2548° (DANIELL); cohesive force of a square inch, when hammered, 33,000 ℔s. (RENNIE).

EARTH, *common.* Specific gravity 1·52 to 2·00; weight of a cubic foot from 95 to 125 ℔s.

ELM. Specific gravity 0·544; weight of a cubic foot 34 ℔s.; weight of a bar 1 foot long and 1 inch square 0·236 ℔s.; will bear on a square inch without permanent alteration 3240 ℔s., and an extension in length of $\frac{1}{414}$; weight of modulus of elasticity for a base of an inch square 1,340,000 ℔s.; height of modulus of elasticity 5,680,000 feet; modulus of

[2] See note to art. 143.—EDITOR.

resilience 7·87; specific resilience 14·4. (Calculated from BARLOW's Experiments.)

Compared with cast iron as unity, its strength is 0·21; its extensibility 2·9; and its stiffness 0·073.

FIR, *red or yellow.* Specific gravity 0·557; weight of a cubic foot 34·8 lbs.; weight of a bar 1 foot long and 1 inch square 0·242 lb.; will bear on a square inch without permanent alteration 4290 lbs., = 2 tons nearly, and an extension in length of $\frac{1}{470}$; weight of modulus of elasticity for a base of an inch square 2,016,000 lbs.; height of modulus of elasticity 8,330,000 feet; modulus of resilience 9·13; specific resilience 16·4. (TREDGOLD.)

Compared with cast iron as unity, its strength is 0·3; its extensibility 2·6; and its stiffness 0·1154, $= \frac{1}{8 \cdot 66}$.

FIR, *white.* Specific gravity 0·47; weight of a cubic foot 29·3 lbs.; weight of a bar 1 foot long and 1 inch square 0·204 lb.; will bear on a square inch without permanent alteration 3630 lbs., and an extension in length of $\frac{1}{504}$; weight of modulus of elasticity for a base of an inch square 1,830,000 lbs.; height of modulus of elasticity 8,970,000 feet; modulus of resilience 7·2; specific resilience 15·3. (TREDGOLD.)

Compared with cast iron as unity, its strength is 0·23; its extensibility 2·4; and its stiffness 0·1.

FLOORS. The weight of a superficial foot of a floor is about 40 lbs. when there is a ceiling, counter-floor, and iron girders. When a floor is covered with people, the load upon a superficial foot may be calculated at 120 lbs. Therefore 120 + 40 = 160 lbs. on a superficial foot is the least stress that ought to be taken in estimating the strength for the parts of a floor of a room.

FORCE. See Gravity, Horses, &c.

GRANITE, *Aberdeen*. Specific gravity 2·625; weight of a cubic foot 164 lbs.; is crushed by a force of 10,910 lbs. upon a square inch. (RENNIE.)

GRAVEL. Weight of a cubic foot about 120 lbs.

GRAVITY, generates a velocity $32\frac{1}{6}$ feet, in a second, in a body falling from rest; space described in the first second $16\frac{1}{12}$ feet.

GUN-METAL, *cast* (copper 8 parts, tin 1). Specific gravity 8·153; weight of a cubic foot $509\frac{1}{2}$ lbs.; weight of a bar 1 foot long and 1 inch square 3·54 lbs. (TREDGOLD); expands in length by 1° of heat $\frac{1}{99090}$ (SMEATON); will bear on a square inch without permanent alteration 10,000 lbs., and an extension in length of $\frac{1}{960}$; weight of modulus of elasticity for a base of an inch square 9,873,000 lbs.; height of modulus of elasticity 2,790,000 feet; modulus of resilience, and specific resilience, not determined (TREDGOLD.)

Compared with cast iron as unity, its strength is 0·65; its extensibility 1·25; and its stiffness 0·535.

HORSE, of average power, produces the greatest effect in drawing a load when exerting a force of $187\frac{1}{2}$ lbs. with a velocity of $2\frac{1}{2}$ feet per second, working 8 hours in a day.[3] (TREDGOLD.) A good horse can exert a force of 480 lbs. for a short time. (DESAGULIERS.) In calculating the strength for horse

---

[3] This is equivalent to raising 3 cubic feet of water $2\frac{1}{2}$ feet per second, or $7\frac{1}{2}$ cubic feet 1 foot per second. See Buchanan's Essays, 3rd edition, by Mr. Rennie, page 88.

machinery, the horse's power should be considered 400 lbs.

IRON, *cast.* See Cast Iron.

IRON, *malleable.* Specific gravity 7·6 (MUSCHENBROËK); weight of a cubic foot 475 lbs.; weight of a bar 1 foot long and 1 inch square 3·3 lbs.; ditto, when hammered, 3·4 lbs.; expands in length by 1° of heat $\frac{1}{143,000}$ (SMEATON); good English iron will bear on a square inch without permanent alteration 17,800 lbs.,[4] = 8 tons nearly, and an extension in length of $\frac{1}{1400}$; cohesive force diminished $\frac{1}{3000}$ by an elevation 1° of temperature; weight of modulus of elasticity for a base of an inch square 24,920,000 lbs.; height of modulus of elasticity 7,550,000 feet; modulus of resilience, and specific resilience, not determined (TREDGOLD).

Compared with cast iron as unity, its strength is 1·12; its extensibility 0·86; and its stiffness 1·3.

LARCH. Specific gravity ·560; weight of a cubic foot 35 lbs.; weight of a bar 1 foot long and 1 inch square 0·243 lb.; will bear on a square inch without permanent alteration 2065 lbs., and an extension in length of $\frac{1}{520}$; weight of modulus of elasticity for a base of an inch square 10,074,000 lbs.; height of modulus of elasticity 4,415,000 feet; modulus of resilience 4; specific resilience 7·1. (Calculated from BARLOW's experiments.)

Compared with cast iron as unity, its strength is 0·136; its extensibility 2·3; and its stiffness 0·058.[5]

LEAD, *cast.* Specific gravity 11·353 (BRISSON); weight

---

[4] Equivalent to a height of 5000 feet of the same matter.

[5] The mean of my trials on specimens of very different qualities

of a cubic foot 709·5 lbs.; weight of a bar 1 foot long and 1 inch square 4·94 lbs.; expands in length by 1° of heat $\frac{1}{62800}$ (SMEATON); melts at 612° (CRICHTON); will bear on a square inch without permanent alteration 1500 lbs., and an extension in length of $\frac{1}{480}$; weight of modulus of elasticity for a base of an inch square 720,000 lbs.; height of modulus of elasticity 146,000 feet; modulus of resilience 3·12; specific resilience 0·27 (TREDGOLD).

Compared with cast iron as unity, its strength is 0·096; its extensibility 2·5; and its stiffness 0·0385.

MAHOGANY, *Honduras*. Specific gravity 0·56; weight of a cubic foot 35 lbs.; weight of a bar 1 foot long and 1 inch square 0·243 lb.; will bear on a square inch without permanent alteration 3800 lbs., and an extension in length of $\frac{1}{420}$; weight of modulus of elasticity for a base of an inch square 1,596,000 lbs.; height of modulus of elasticity 6,570,000 feet; modulus of resilience 9·047; specific resilience 16·1. (TREDGOLD.)

Compared with cast iron as unity, its strength is 0·24; its extensibility 2·9; and its stiffness 0·487.

MAN. A man of average power produces the greatest effect when exerting a force of 31¼ lbs. with a velocity of 2 feet per second, for 10 hours in a day.[6] (TREDGOLD.) A strong man will raise and carry from 250 to 300 lbs. (DESAGULIERS.)

---

places the strength and stiffness of Larch much higher on the scale of comparison; but I had not observed the point where it loses elastic power.

[6] This is equivalent to half a cubic foot of water raised 2 feet per second; or 1 cubic foot of water 1 foot per second. See Buchanan's Essays, p. 92, edited by Mr. Rennie.

MARBLE, *white.* Specific gravity 2·706; weight of a cubic foot 169 lbs.; weight of a bar 1 foot long and 1 inch square 1·17 lbs.; cohesive force of a square inch 1811 lbs.;[7] extensibility $\frac{1}{1394}$ of its length; weight of modulus of elasticity for a base of an inch square 2,520,000 lbs.; height of modulus of elasticity 2,150,000 feet; modulus of resilience at the point of fracture 1·3; specific resilience at the point of fracture 0·48 (TREDGOLD); is crushed by a force of of 6060 lbs. upon a square inch (RENNIE).

MERCURY. Specific gravity 13·568 (BRISSON); weight of a cubic inch 0·4948 lb.; expands in bulk by 1° of heat $\frac{1}{9990}$ (DULONG and PETIT); weight of modulus of elasticity for a base of an inch square 4,417,000 lbs.; height of modulus of elasticity 750,000 feet (Dr. YOUNG, from CANTON's Experiments).

OAK, *good English.* Specific gravity 0·83; weight of a cubic foot 52 lbs.; weight of a bar 1 foot long and 1 inch square 0·36 lb.; will bear upon a square inch without permanent alteration 3960 lbs., and an extension in length of $\frac{1}{430}$; weight of modulus of elasticity for a base of an inch square 1,700,000 lbs.; height of modulus of elasticity 4,730,000 feet; modulus of resilience 9·2; specific resilience 11. (TREDGOLD.)

Compared with cast iron as unity, its strength is 0·25; its extensibility 2·8; and its stiffness 0·093.

---

[7] My experiments give 551 lbs. for the cohesive strength per square inch of white marble. The value 1811 was in this, as in other cases, calculated by Tredgold from the same erroneous supposition that he obtained the great strength of cast iron, art. 143.—EDITOR.

PENDULUM. Length of pendulum to vibrate seconds in the latitude of London 39·1372 inches (KATER); ditto to vibrate half seconds 9·7843 inches.

PINE, *American, yellow.* Specific gravity 0·46; weight of a cubic foot 26¾ lbs.; weight of a bar 1 foot long and 1 inch square 0·186 lb.; will bear on a square inch without permanent alteration 3900 lbs., and an extension in length of $\frac{1}{414}$; weight of modulus of elasticity for a base of an inch square 1,600,000 lbs.; height of modulus of elasticity 8,700,000 feet; modulus of resilience 9·4; specific resilience 20. (TREDGOLD.)

Compared with cast iron as unity, its strength is 0·25; its extensibility 2·9; and its stiffness 0·087.

PORPHYRY, *red.* Specific gravity 2·871; weight of a cubic foot 179 lbs.; is crushed by a force of 35,568 lbs. upon a square inch. (GAUTHEY.)

ROPE, *hempen.* Weight of a common rope 1 foot long and 1 inch in circumference from 0·04 to 0·046 lb.; and a rope of this size should not be exposed to a strain greater than 200 lbs.; but in compounded ropes, such as cables, the greatest strain should not exceed 120 lbs.;[8] and the weight of a cable 1 foot in length and 1 inch in circumference does not exceed 0·027 lb.

---

[8] The square of the circumference in inches multiplied by 200 will give the number of pounds a rope may be loaded with, and multiply by 120 instead of 200 for cables. Common ropes will bear a greater load with safety after they have been some time in use, in consequence of the tension of the fibres becoming equalized by repeated stretchings and partial untwisting. It has been imagined that the improved strength was gained by their being laid up in store; but if they can there be preserved from deterioration, it is as much as can be expected.

ROOFS. Weight of a square foot of Welsh rag slating 11¼ lbs.; weight of a square foot of plain tiling 16¼ lbs.; greatest force of the wind upon a superficial foot of roofing may be estimated at 40 lbs.

SLATE, *Welsh.* Specific gravity 2·752 (KIRWAN); weight of a cubic foot 172 lbs.; weight of a bar 1 foot long and 1 inch square 1·19 lbs.; cohesive force of a square inch 11,500 lbs.; extension before fracture $\frac{1}{1370}$; weight of modulus of elasticity for a base of an inch square 15,800,000 lbs.; height of modulus of elasticity 13,240,000 feet; modulus of resilience 8·4; specific resilience 2 (TREDGOLD.)

SLATE, *Westmoreland.* Cohesive force of a square inch 7870 lbs.; extension in length before fracture $\frac{1}{1640}$; weight of modulus of elasticity for a base of an inch square 12,900,000 lbs. (TREDGOLD.)

SLATE, *Scotch.* Cohesive force of a square inch 9600 lbs.; extension in length before fracture $\frac{1}{1645}$; weight of modulus of elasticity for a base of an inch square 15,790,000 lbs. (TREDGOLD.)

STEAM. Specific gravity at 212° is to that of air at the mean temperature as 0·472 is to 1 (THOMSON); weight of a cubic foot 249 grains; modulus of elasticity for a base of an inch square 14¾ lbs.; when not in contact with water, expands $\frac{1}{480}$ of its bulk by 1° of heat (GAY LUSSAC.)

STEEL. Specific gravity 7·84; weight of a cubic foot 490 lbs.; a bar 1 foot long and 1 inch square weighs 3·4 lbs.; it expands in length by 1° of heat $\frac{1}{157200}$ (ROY); tempered steel will bear without permanent alteration 45,000 lbs.; cohesive force of a square inch 130,000 lbs. (RENNIE); cohesive force diminished

$\frac{1}{5000}$ by elevating the temperature 1°; modulus of elasticity for a base of an inch square 29,000,000 lbs.; height of modulus of elasticity 8,530,000 feet (Dr. YOUNG.)

STONE, *Portland.* Specific gravity 2·113; weight of a cubic foot 132 lbs.; weight of a prism 1 inch square and 1 foot long 0·92 lb.; absorbs $\frac{1}{16}$ of its weight of water (R. TREDGOLD); is crushed by a force of 3729 lbs. upon a square inch (RENNIE); cohesive force of a square inch 857 lbs.; extends before fracture $\frac{1}{1789}$ of its length; modulus of elasticity for a base of an inch square 1,533,000 lbs.; height of modulus of elasticity 1,672,000 feet;[9] modulus of resilience at the point of fracture 0·5; specific resilience at the point of fracture 0·23 (TREDGOLD.)

STONE, *Bath.* Specific gravity 1·975; weight of a cubic foot 123·4 lbs.; absorbs $\frac{1}{13}$ of its weight of water (R. TREDGOLD); cohesive force of a square inch 478 lbs. (TREDGOLD.)

STONE, *Craigleith.* Specific gravity 2·362; weight of a cubic foot 147·6 lbs.; absorbs $\frac{1}{63}$ of its weight of water; cohesive force of a square inch 772 lbs. (TREDGOLD); is crushed by a force of 5490 lbs. upon a square inch (RENNIE.)

STONE, *Dundee.* Specific gravity 2·621; weight of a cubic foot 163·8 lbs.; absorbs $\frac{1}{511}$ part of its weight of water; cohesive force of a square inch 2661 lbs.

---

[9] In the stones, the modulus here given is calculated from the flexure at the time of fracture; when it is taken for the first degrees of flexure, it is a little greater. The experiments are described in the Philosophical Magazine, vol. lvi. p. 290.

(TREDGOLD); is crushed by a force of 6630 lbs. upon a square inch (RENNIE.)

STONE-WORK. Weight of a cubic foot of rubble-work about 140 lbs.; of hewn stone 160 lbs.

TIN, *cast*. Specific gravity 7·291 (BRISSON); weight of a cubic foot 455·7 lbs.; weight of a bar 1 foot long and 1 inch square 3·165 lbs.; expands in length by 1° of heat $\frac{1}{72510}$ (SMEATON); melts at 442° (CRICHTON); will bear upon a square inch without permanent alteration 2880 lbs., and an extension in length of $\frac{1}{1600}$; modulus of elasticity for a base of an inch square 4,608,000 lbs.; height of modulus of elasticity 1,453,000 feet; modulus of resilience 1·8; specific resilience 0·247 (TREDGOLD.)

Compared with cast iron as unity, its strength is 0·182; its extensibility 0·75; and its stiffness 0·25.

WATER, *river*. Specific gravity 1·000; weight of a cubic foot 62·5 lbs.; weight of a cubic inch 252·525 grains; weight of a prism 1 foot long and 1 inch square 0·434 lb.; weight of an ale gallon of water 10·2 lbs.; expands in bulk by 1° of heat $\frac{1}{3858}$ (DALTON);[10] expands in freezing $\frac{1}{17}$ of its bulk (WILLIAMS); and the expanding force of freezing water is about 35,000 lbs. upon a square inch, according to Muschenbroëk's valuation; modulus of elasticity for a base of an inch square 326,000 lbs.; height of mo-

[10] Water has a state of maximum density at or near 40°, which is considered an exception to the general law of expansion by heat: it is extremely improbable that there is any thing more than an apparent exception, most likely arising from water at low temperatures absorbing a considerable quantity of air, which has the effect of expanding it; and consequently of causing the apparent anomaly.

dulus of elasticity 750,000 feet, or 22,100 atmospheres of 30 inches of mercury (Dr. YOUNG, from CANTON's Experiments).

WATER, *sea*. Specific gravity 1·0271; weight of a cubic foot 64·2 lbs.

WATER is 828 times the density of air of the temperature 60°, and barometer 30.

WHALE-BONE. Specific gravity 1·3; weight of a cubic foot 81 lbs.; will bear a strain of 5600 lbs. upon a square inch without permanent alteration, and an extension in length of $\frac{1}{146}$; modulus of elasticity for a base of an inch square 820,000 lbs.; height of modulus of elasticity 1,458,000 feet; modulus of resilience 38·3; specific resilience 29. (TREDGOLD.)

WIND. Greatest observed velocity 159 feet per second (ROCHON); force of wind with that velocity about 57¾ lbs. on a square foot.[11]

---

[11] Table of the force of winds, formed from the Tables of Mr. Rouse and Dr. Lind, and compared with the observations of Colonel Beaufoy.

Velocity in miles per hour.	A wind may be denominated when it does not exceed the velocity opposite to it.	Velocity per second.	Force on a square foot.
		feet.	lbs.
6·8	A gentle pleasant wind . . .	10	0·229
13·6	A brisk gale	20	0·915
19·5	A very brisk gale	30	2·059
34·1	A high wind	50	5·718
47·7	A very high wind	70	11·207
54·5	A storm or tempest	80	14·638
68·2	A great storm	100	22·872
81·8	A hurricane	120	32·926
102·3	A violent hurricane, that tears up trees, overturns buildings, &c.	150	51·426

Accurate observations on the variation and mean intensity of

ZINC, *cast*. Specific gravity 7·028 (WATSON); weight of a cubic foot 439¼ lbs.; weight of a bar 1 inch square and 1 foot long 3·05 lbs.; expands in length by 1° of heat $\frac{1}{61200}$ (SMEATON); melts at 648° (DANIELL); will bear on a square inch without permanent alteration 5700 lbs. = 0·365 cast iron, and an extension in length of $\frac{1}{2400}$ = ½ that of cast iron (TREDGOLD);[12] modulus of elasticity for a base of an inch square 13,680,000 lbs.; height of modulus of elasticity 4,480,000 feet; modulus of resilience 2·4; specific resilience 0 34. (TREDGOLD.)

Compared with cast iron as unity, its strength is 0·365; its extensibility 0·5; and its stiffness 0·76.

---

the force of winds would be very desirable both to the mechanician and meteorologist.

[12] The fracture of zinc is very beautiful; it is radiated, and preserves its lustre a long time.

NOTE ON THE ACTION OF CERTAIN SUBSTANCES ON CAST IRON.

In some circumstances cast iron will decompose, and be converted into a soft substance resembling plumbago. A few instances of this kind I add here, as they will be interesting to persons who employ iron for various purposes.

Dr. Henry observed that when cast iron was left in contact with muriate of lime, or muriate of magnesia, most of the iron was removed, the specific gravity of the mass was reduced to 2·155, and what remained consisted chiefly of plumbago, and the impurities usually found in cast iron.[1]

A similar change was produced in some cast iron cylinders used to apply the weaver's dressing to cloth: this dressing is a kind of paste, made of wheat or barley-meal. The corrosion of the cylinders took place repeatedly, and was so rapid that it was found necessary to use wooden ones. Dr. Thomson ascribes the change to the acid formed by the paste turning sour.[2]

Another instance of greater importance has been recorded by Mr. Brande. A portion of a cast iron gun had undergone a like change from being long immersed in sea-water. To the depth of an inch it was converted into a substance having all the external characters of plumbago; it was easily cut, greasy to the feel, and made a black streak upon paper.[3] The component parts of this substance were in the ratio of

---

[1] Dr. Thomson's Annals of Philosophy, vol. v. p. 66.

[2] Idem, vol. x. p. 302.

[3] Quarterly Journal of Science, vol. xii. p. 407.

Oxide of iron	81
Plumbago	16
	97

Mr. Brande could not detect any silica in it; and remarks, that anchors and other articles of wrought iron, when similarly exposed, are only superficially oxidized, and exhibit no other peculiar appearance.

Near the town of Newhaven, in America, a cannon ball was discovered, which it was ascertained had lain undisturbed about forty-two years in ground kept constantly moist by sea-water: the diameter of the ball was 3·87 inches, and with a common saw a section was easily made through a coat of plumbaginous matter, which at the place of incision was half an inch thick; but its thickness varied in different places. The plumbago cut with the same ease, gave the same streak to paper, and had in every respect the properties of common black lead.

A cannon ball had undergone a similar change, which was taken from the wreck of a vessel that appeared to have been many years under water: the ball was covered by oysters firmly adhering to it, and its external part was converted into plumbago. But an old cannon found covered with oysters did not, on the removal of its coating, show any signs of such conversion.[4]

The reader who wishes to pursue this interesting subject may consult an article "On the Mechanical Structure of Iron developed by Solution," &c., by Mr. Daniell,[5] who has made several experiments with a view to determine the nature of the substance resembling plumbago, which is found on the surface of iron after it has been exposed to the action of an acid.

---

[4] Phillips's Annals of Philosophy, vol iv. p. 77. 1822.

[5] Quarterly Journal of Science, vol. ii. p. 278. Much additional information on the effect of water upon iron will be obtained from the Report of Mr. Mallet "upon the Action of Sea and River Water, whether clear or foul, and at various temperatures, upon Cast and Wrought Iron." British Association, vols. vii. and viii.—Editor.

Mr. Daniell found that the structure of iron, as developed by solution, was very different in different kinds; and that it required three times as long to saturate a given portion of acid when it acted on white cast iron, as when it acted on the gray kind.

PLATE I.

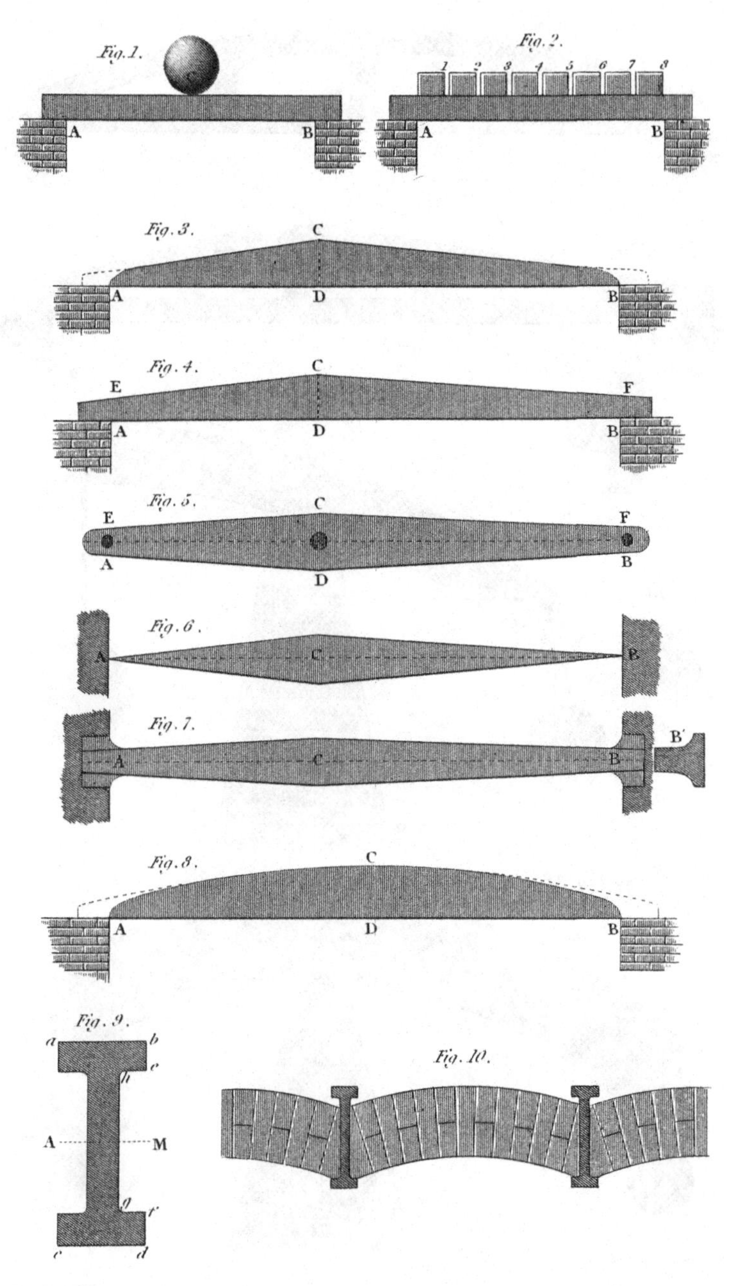

…wn by Thos. Tredgold.

Published by J. Weale, at the Architectural Library, 59, High Holborn, 1842.

PLATE II.

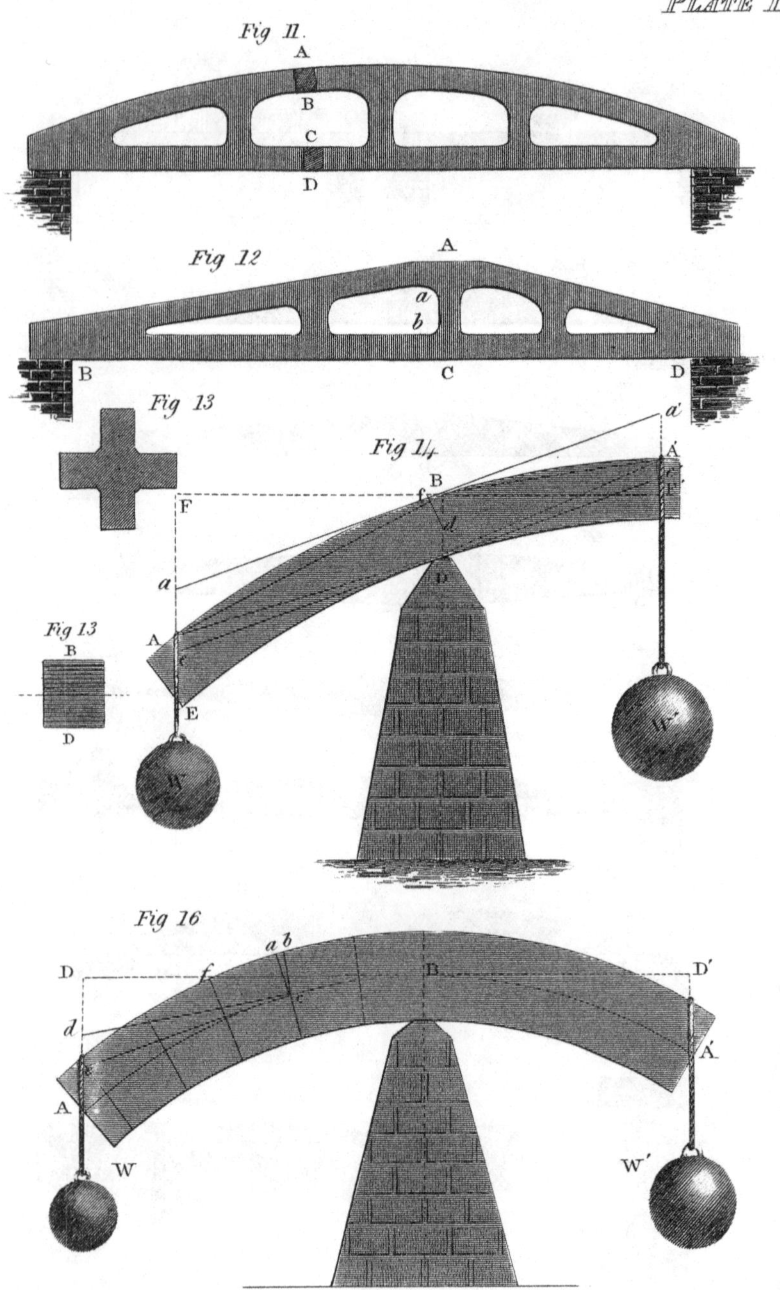

Drawn by Thos. Tredgold. *Engraved by R.W.S.*

Published by J. Weale, at the Architectural Library, 59, High Holborn, 1842.

PLATE III.

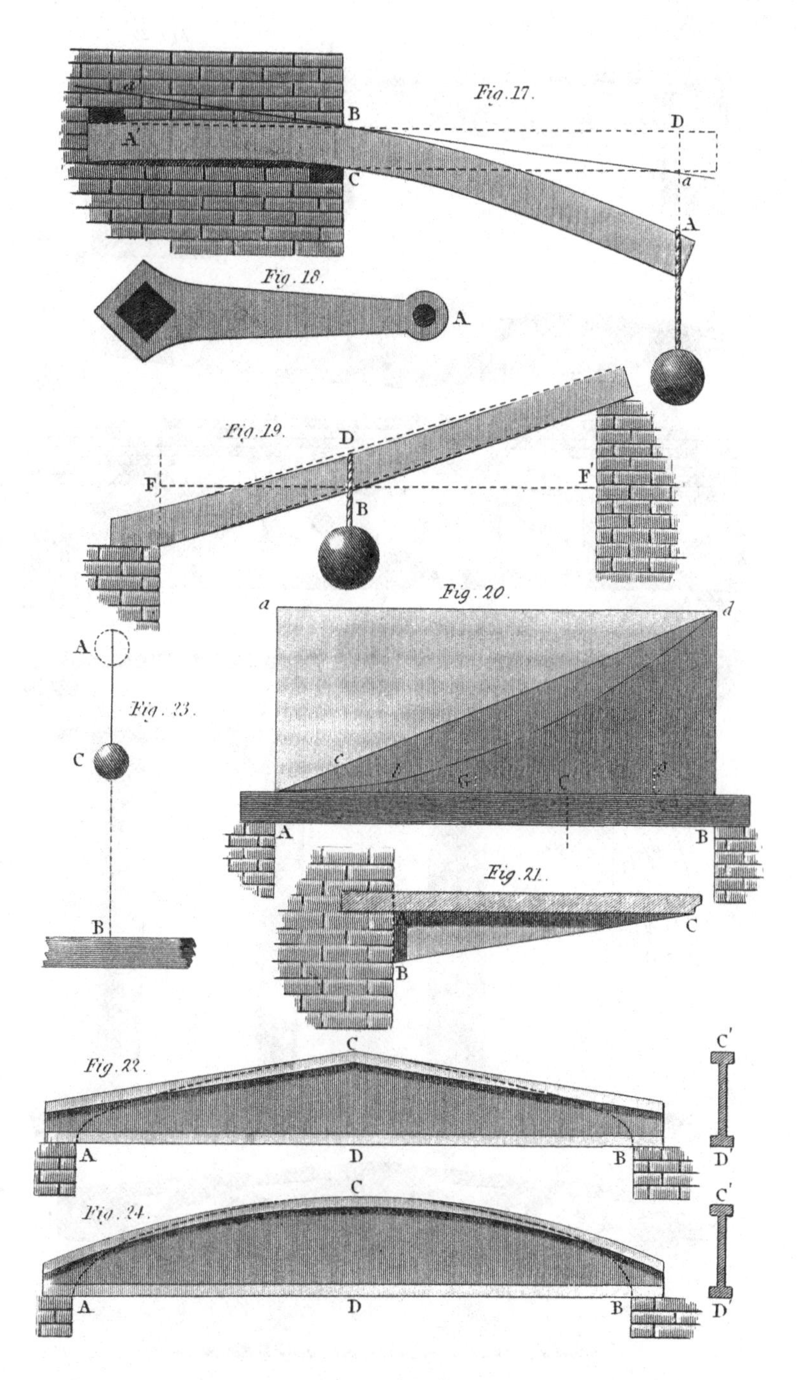

Published by J. Weale, at the Architectural Library, 59, High Holborn, 1842.

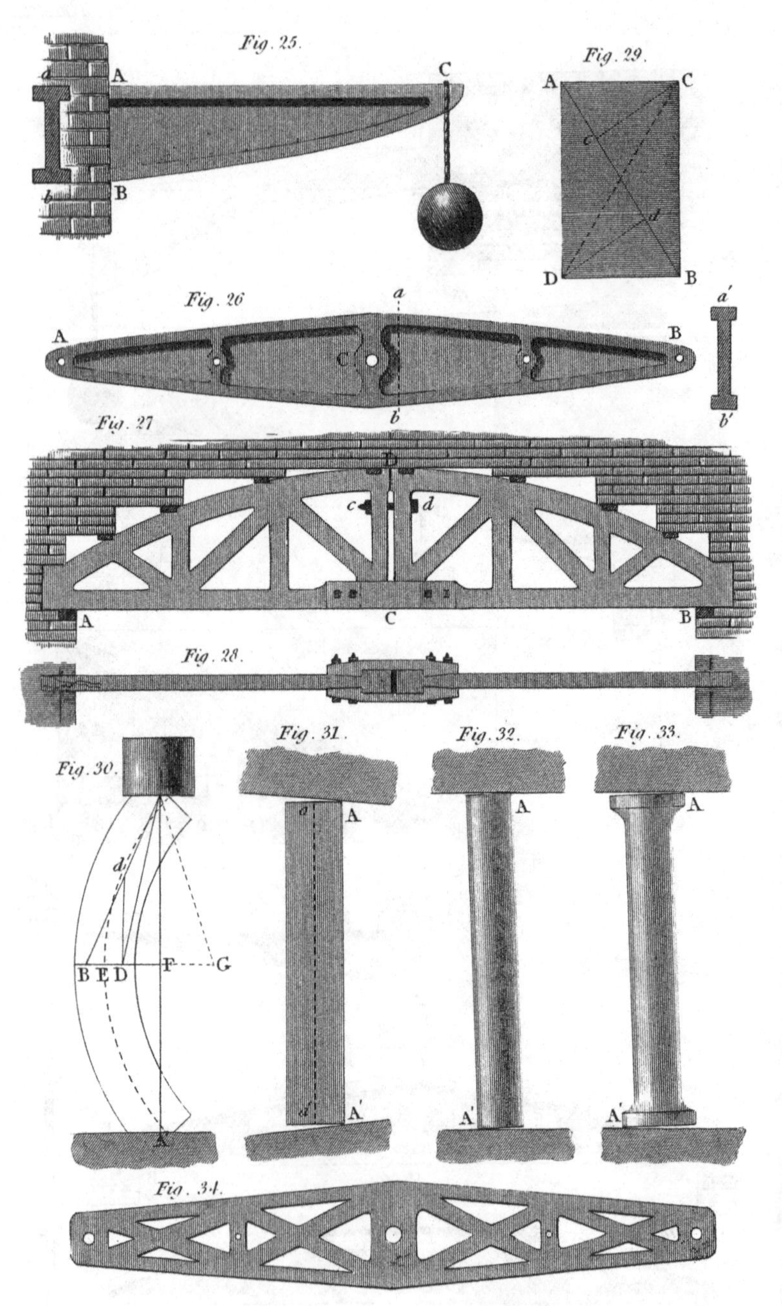

Published by J. Weale, at the Architectural Library, 59, High Holborn, 1842.

EXPLANATION OF THE PLATES.

PLATE I.

FIG. 1. A bar supported at the ends, and loaded in the middle of the length. See art. 8.

FIG. 2. A beam with the load uniformly distributed over the length, as the experiment, art. 61, was tried. See art. 20 and 61.

FIG. 3. The form for a beam of uniform strength to resist the action of a load at C. ACD and BCD are semi-parabolas, A and B being the vertices. The dotted lines show the additions to this form to render it of practical use. See art. 27, 123, and 223-229.

FIG. 4. A form for a beam which is as nearly of uniform resistance as practical conditions will admit of: it is bounded by straight lines, and the depths at the ends are each half the depth in the middle. See art. 17, (Ex. 7,) 28, 65, 127, and 230-234.

FIG. 5. A variation of the last form for the case where the force sometimes acts upwards and sometimes downwards. See art. 29, 127, and 230-234.

FIG. 6. A figure of uniform strength for a beam, when the depth is uniform. See art. 30, 122, and 242-246.

FIG. 7. A modification of fig. 6, which is the most economical form of equal strength for resistance to pressure. B′ is the form of the end. See art. 30.

FIG. 8. The form of equal resistance for a load rolling along the upper side, as in a railway; or for a load uniformly distributed over the length. ACB is half an ellipse. The dotted lines show the addition required in practice. See art. 32, 125, 240, and 241.

FIG. 9. The strongest form of section for a beam to resist a cross strain. AM is the line called the *neutral axis*. See art. 40, 54, 116, 185-197, and 321.

FIG. 10 shows an application of the section, fig. 9, to form a fire-proof floor, the projection serving the double purpose of giving additional strength, and forming a support for the arches. See art. 40 and 194.

PLATE II.

FIG. 11. This is the figure of a very economical beam for supporting a load diffused over its length: it is adapted for girders, beams to support walls, and the like. An easy rule for proportioning girders is given in art. 50. When this form is used for a girder, the openings answer for inserting cross joists. AB and CD show the sections at these places. See art. 21, (Ex. 12,) 41, 43, 117, 198-210, and 323.

FIG. 12. This figure shows a beam on the same principle as the preceding figure, except that the load is supposed to act only at one point A. See art. 43, 44, 117, and 198-210.

FIG. 13. The section of a shaft, commonly called a feathered shaft. See art. 46.

FIG. 14. A figure to illustrate the action of forces upon a beam, and to explain the mode of calculation. See art. 106, 108, 131, and 154.

FIG. 15. A section of the beam in fig. 14, at BD. This section is supposed to be divided into thin laminæ. See art. 106.

FIG. 16. A figure to illustrate the method of calculating the deflexion of beams. In these figures (figs. 14 and 16) I have regarded distinctness of the parts referred to, more than the true relation of the parts to one another. See art. 120.

PLATE III.

FIG. 17 is to illustrate the circumstances which take place in the deflexion of beams fixed at one end. See art. 133 and 154.

FIG. 18. To explain the mode of calculating the strength of cranks. See art. 135.

FIG. 19. A figure to explain the manner of estimating the strength and deflexion of a beam supported at the ends. See art. 136, 143, 146, 149, and 165.

FIG. 20. A figure to show how to calculate the strain upon a beam when a load is distributed in any regular manner over it. The load being uniform, *a d* is the line which represents its upper surface; when the load increases as the distance from the end A, *c d* is the line bounding it; and when the load increases as the square of the distance from A, *b d* is the line bounding it. The second case is the same as the pressure of a fluid against a vertical sheet fixed at the ends. See art. 138-141, and 160.

FIG. 21. ACB is the form of equal strength for an uniform load: it is in this figure applied to the cantilever of a balcony, and whatever ornamental form may be given to the under side, it should not be cut within the line B C. See art. 34, 130, and 157.

FIG. 22. When the section is of the form C′ D′, and the breadth uniform, the figure of equal strength for a load in the middle is formed by two semi-parabolas, (as in fig. 3,) shown by the dotted lines; and it may be formed for practical application, as shown in the figure. See art. 187, 223-227.

FIG. 23 is a figure to illustrate the nature of variable forces. See art. 295.

FIG. 24. If the section of a beam be C′ D′, and the breadth uniform, the form of equal strength for an uniform load, as in girders for floors, is a semi-ellipse, shown by the dotted lines; and also when the load rolls or slides over it; and it may be formed for practical application, as the figure; and an easy rule for girders of this kind is given in page 42. See art. 21, (Ex. 13,) 188, 193, 240, 241, and 322.

PLATE IV.

FIG. 25 represents a beam fixed at one end; $a' b'$ is its section; the load acting at the end C, the figure of equal resistance is a semi-parabola. See art. 196 and 223-229.

FIG. 26. Form suitable for the beam of a steam engine to the form of section $a' b'$. See art. 40, 196, 222, and 223-229.

FIG. 27. A sketch for a beam to bear a considerable load distributed uniformly over its length, when the span is so much as to render it necessary to cast it in two pieces. The connexion may be made by a plate of wrought iron on each side at C, with indents to fit the corresponding parts of the beam. Wrought iron should be preferred for the connecting plates, because, being more ductile, it is more safe. See the next figure and art. 198-210.

FIG. 28 shows the under side of the beam in the preceding figure. The plates are held together by bolts; but it is intended that the strength should depend on the incidents, the bolts being only to hold them together. No connexion is wanted at the upper side of the beam, except a bolt $c\ d$, or like contrivance, to steady it. See art. 199.

FIG. 29. A figure to explain the nature of the resistance to twisting or torsion. See art. 263.

FIG. 30. A figure to illustrate the action of the straining force on columns, posts, and the like. See art. 276.

FIG. 31. To explain the effect of settlement or other derangement of the straining force. See art. 10 and 281.

FIG. 32. Another case of settlement or derangement of the straining force on a column considered. See art. 283.

Fig. 33. To show why columns should not be enlarged at the top or bottom. See art. 283.

Fig. 34. A sketch for an open beam recommended for an engine beam. See art. 350. In small beams the middle part may be left wholly open, except at the centre. Capt. Kater has used this form for the beam of a delicate balance.

A LIST OF AUTHORS

QUOTED IN THE ALPHABETICAL TABLE, WITH THE TITLES OF THE WORKS FROM WHICH THE DATA HAVE BEEN QUOTED.

BARLOW. Essay on the Strength and Stress of Timber, London, 1817; reprinted with many additions in 1837.

BEAUFOY. Thomson's Annals of Philosophy.

BRISSON. Dr. Thomson's System of Chemistry, fifth edition, London, 1817.

CRICHTON. Philosophical Magazine, vol. xv.

DALTON. Dr. Thomson's System of Chemistry, fifth edit.

DANIELL. Quarterly Journal of Science, vol. xi. p. 318.

DESAGULIERS. Course of Experimental Philosophy, London, 1734.

DULONG and PETIT. Annals of Philosophy, vol. xiii.

GAUTHEY. Rozier's Journal de Physique, tome iv. p. 406.

GAY LUSSAC. Dr. Thomson's System of Chemistry, fifth edit.

HATCHETT. Dr. Thomson's System of Chemistry, fifth edit.

KATER. Philosophical Magazine, vol. liii., 1818.

KIRWAN. Elements of Mineralogy.

LIND. Dr. Thomas Young's Lectures on Natural Philosophy.

MUSCHET. Philosophical Magazine, vol. xviii.

MUSCHENBROËK. Dr. Thomson's System of Chemistry, fifth edit.

RENNIE. Philosophical Transactions for 1818, Part I.

RICE. Annals of Philosophy for 1819.

ROCHON. Dr. Thomas Young's Natural Philosophy, vol. ii.

ROUSE. Smeaton's Experimental Inquiry of the Power of Wind and Water.

ROY. Account of the Trigonometrical Survey of England and Wales, vol. i.

ROYAL SOCIETY. Dr. Thomas Young's Natural Philosophy, vol. ii.

SHUCKBURGH. Dr. Thomson's System of Chemistry.

SMEATON. Reports, vol. iii. and Miscellaneous Papers.

THOMSON. Annals of Philosophy, vol. iii., New Series, 1822.

R. TREDGOLD. Elementary Principles of Carpentry, London, 1820, of which an enlarged edition has been given by Mr. Barlow in 1840.

TREDGOLD. Philosophical Magazine, vol. lvi., 1820; Elementary Principles of Carpentry; Buchanan's Essays on Mill-work, second edit., or third edit. by Mr. Rennie in 1841; and Experiments of which the details have not been published.

TROUGHTON. Dr. Thomas Young's Natural Philosophy, vol. ii.

WATSON. Chemical Essays.

WILLIAMS. Dr. Thomas Young's Natural Philosophy, vol. ii.

YOUNG. Natural Philosophy, vol. ii.

Printed in the United States
By Bookmasters